T0340190

Guide to Cell Therapy GxP

Guide to Cell Therapy GxP

Quality Standards in the Development of Cell-Based Medicines in Non-Pharmaceutical Environments

Joaquim Vives
Barcelona, Spain

Gloria Carmona
Seville, Spain

ELSEVIER

AMSTERDAM • BOSTON • HEIDELBERG • LONDON
NEW YORK • OXFORD • PARIS • SAN DIEGO
SAN FRANCISCO • SINGAPORE • SYDNEY • TOKYO
Academic Press is an imprint of Elsevier

Academic Press is an imprint of Elsevier
125 London Wall, London EC2Y 5AS, UK
525 B Street, Suite 1800, San Diego, CA 92101-4495, USA
225 Wyman Street, Waltham, MA 02451, USA
The Boulevard, Langford Lane, Kidlington, Oxford OX5 1GB, UK

ISBN: 978-0-12-803115-5

British Library Cataloguing-in-Publication Data
A catalog record for this book is available from the British Library

Library of Congress Cataloging-in-Publication Data
A catalog record for this book is available from the Library of Congress

For information on all Academic Press publications
visit our website at http://store.elsevier.com/

Working together
to grow libraries in
developing countries

www.elsevier.com • www.bookaid.org

Acquisition Editor: Mica Haley
Editorial Project Manager: Lisa Eppich
Production Project Manager: Karen East and Kirsty Halterman
Designer: Greg Harris

Typeset by TNQ Books and Journals
www.tnq.co.in

Printed and bound in the United States of America

Contents

List of Contributors

Jacqueline Barry Cell Therapy Catapult Ltd, Guys Hospital, London, UK

Ana Cardesa Andalusian Initiative for Advanced Therapies, Consejeria de Salud. Junta de Andalucia, Sevilla, Spain

Natividad Cuende Andalusian Initiative for Advanced Therapies. Servicio Andaluz de Salud, Consejería de Salud, Junta de Andalucía. Sevilla. Spain

Ander Izeta Tissue Engineering Laboratory, Bioengineering Area, Instituto Biodonostia, Hospital Universitario Donostia, San Sebastian, Spain

Giulia Leoni Cell Therapy Catapult Ltd, Guys Hospital, London, UK

Fabiola Lora Andalusian Initiative for Advanced Therapies, Consejeria de Salud. Junta de Andalucia, Sevilla, Spain

Rosario C. Mata Andalusian Initiative for Advanced Therapies, Consejeria de Salud. Junta de Andalucia, Sevilla, Spain

Blanca Miranda Biobank Andalousian Public Health Care System Granada, Spain

Natalie Mount Cell Therapy Catapult Ltd, Guys Hospital, London, UK

Roger Palau Banc de Sang i Teixits, Edifici Dr. Frederic Duran i Jordà, Passeig Taulat, Barcelona, Spain

Arnau Pla CELLAB, Barcelona, Spain

Andy Römhild Berlin-Brandenburg Center for Regenerative Therapies (BCRT), Charité University Medicine Berlin, Berlin, Germany

Rindi Schutte Cell Therapy Catapult Ltd, Guys Hospital, London, UK

Michaela Sharpe Cell Therapy Catapult Ltd, Guys Hospital, London, UK

Amy Lynnette Van Deusen University of Virginia, Department of Pharmacology, Charlottesville, VA, USA

Foreword

This book represents an essential guide capable of presenting and dissecting the steps of a revolution that has been taking place in cell manufacturing with a specific but not exclusive focus on the so-called hospital-based facilities.

The introduction of more stringent rules in cell-based therapeutics development started just after the mid-2000 in Europe and in other regions of the globe. This mandated new operational rules and new ways of thinking. A sort of "industrial revolution" in cellular therapy aimed to change the manner in which a cell-based treatment had to be conceived, manufactured, and introduced into different clinical scenarios. It was a true challenge for both the producers and the regulators.

The lines in this precious text are taking advantage of almost 10 years of these complex evolutions and experience, accompanying the reader within proper product developmental strategies that consider operational contaminations between enterprises and hospital facilities as a way to generate quality assurance systems capable of facing new regulations and providing better products in early and late clinical phases.

This "gmp-ification" transition was relatively easily absorbed by industries, finding more resistance in less prepared hospital environments. The authors address this aspect, outlining which ways hospital-based facilities were able to face this "industrial revolution" in cell manufacturing. This phase also overlapped with an unprecedented crisis in world economy investing public institutions and enterprises and negatively impacting the investments for novel therapeutic approaches, cells included.

Here, the authors give evidences on how, despite the crisis, these two words were able to adapt and share their strategies to satisfy regulatory requirements, producing a quality system for GxP implementation that is now generating products with solid legs able to walk the different paths of human diseases.

The reader is lead to a road map for product development from pre-clinical evidences to clinical translations, depicting regulatory framework in Europe with clarification on definitions and regulations for the so-called Advanced Therapeutics Medicinal Product (ATMP) manufacturing. While mostly focused on the European context, the authors attempt to reflect their guidance onto other frameworks, suggesting a need for harmonization within different regulations.

The several phases of a cell-based therapy development are described with enlightening examples. Indications of the need for preclinical strategies to assess efficacy and safety are provided—outlining how all these steps should be cell specific and disease related. Instructions on how to design nonclinical programs to valorize the power of a cell-based product and to address concerns in a risk-based approach are described in a simple but not reductive manner. All these aspects will always find a partner in the International Society for Cellular Therapy (ISCT), which I am currently chairing.

The entire book is a Good X Practice analysis, where the "X-factors" are represented by laboratory (L), manufacturing (M), and clinical (C) steps that should take a scientifically sound concept and translate it for patient care. In particular, one can appreciate the proposal of a consistent introduction of clinical research approaches to assess the safety and the potential of a cellular product abandoning passionate expectations to look for measurable end-points within appropriate follow-up.

My feeling is that the editors and the authors of this book have truly centered on hot topics in cellular therapeutics have collected milestone contributions. I additionally have the vivid sensation that this book will play a relevant role in educating academia and industry. Many clinicians, scientists, technicians, developers, and all cellular therapists should be grateful to these authors who were here able to share their pioneering experience in the field.

Massimo Dominici, MD
University Hospital of Modena and Reggio Emilia, Italy
President of the International Society for Cellular Therapy 2014–2016
Modena, April, 13 2015

Preface

The possibility of treating diseases using complete organs, tissues, or isolated cells from the very same patient or a suitable donor has been, and still is, one of the challenges that has spurred biomedical research in recent decades. Availability, immunotolerance, engraftment rate and long-term full recovery of function remain active battles.

From a historical perspective, transfusion medicine has been the first known strategy that attempted to treat patients with live cells. Since the first description of blood groups until the modern manufacturing and molecular typing, transfusion has dramatically evolved at the scientific and therapeutic levels. Millions of people today owe their lives to voluntary donors and to the current sophisticated network of blood services (https://www.ibms.org/go/nm:history-blood-transfusion).

Going one step further, the understanding that cellular therapies involve activities aiming to the long-term repair or substitution of damaged cells or tissues, it can be stated without a doubt that hematopoietic transplantation is the oldest and more widespread form of cellular therapy.

Hematopoietic transplantation was born and evolved through the early experiences of George Mathé [1], followed by the recognition of histo-compatibility antigens [2], the actual key of the current transplants universe. Finally, the systematization and positive results achieved by the Seattle group that ultimately won the Nobel Prize for ED Thomas [3] were the basis for current transplant strategies, which are the only solution for a number of lympho-hemopoietic disorders. It was the first demonstration that some cells could cure or help to cure diseases. Since the 70s, more than 40 years of research and clinical investigation have confirmed the expectations of these treatments through their different evolutions. These include not only allogeneic matched and mismatched sibling transplants but also transplants using autologous bone marrow and peripheral blood stem cells, or unrelated bone marrow or cord blood, which in a number of cases are receiving ex vivo modified products.

In parallel, research in cellular, molecular, and developmental biology (especially that on stem cells of different origin and evolutionary status) has triggered the spreading of new therapeutic approaches in the form of immunotherapy [4], gene therapy, and more recently, cellular therapies aimed at regeneration of tissues [5] and perhaps organs, consolidating a new area of complex knowledge: regenerative medicine [6].

Regenerative medicine, based on the use of cells as medicines, is evolving very intensely, creating new therapeutic opportunities but also social and economic value at a rate that has not been recently observed in other areas of science. With the conviction that cellular therapies could benefit millions of citizens, the European Union and many countries are creating specific programs and structures to promote new developments and treatments. The envisaged potential huge market has stimulated private investment in hundreds of companies, mostly new, which has duplicated in the last two years, generating an overall growth of economic value of more than 25% per year [7].

Besides scientific development, new effective treatments, and value creation, regulators and scientific societies have progressively set up a frame of laws and quality assurance systems looking for something that is inalienable: the quality and safety of products and treatments.

Again, blood transfusion has pioneered the implementation of good practices standards. The American Association of Blood Banks (AABB) released its first blood bank standards in 1957 [8], which were followed in the 80' by standards for Cellular Therapies, Perioperative Services, Relationship Testing, Immunohematology Reference Laboratories, Molecular Testing, and Patient Blood Management being added to the standards library (www.aabb.org).

In addition, specialized societies and organizations, such as the International Society for Cellular Therapy (ISCT, www.celltherapysociety.org) and Netcord (www.netcord.org) in collaboration with the Foundation for the Accreditation of Cellular Therapy (FACT, www.factwebsite.org), have developed international standards (FACT/JACIE, FACT/Netcord) and implemented accreditation procedures. In a number of countries, local standards and other authorization/certification and/or accreditation procedures are also in place with the common aim of ensuring donor and patient safety.

More recently, the definition of a new therapeutic category within the cell therapy area, for substantially manipulated cell-based medicines, has promoted a new regulatory framework that has consolidated into specific regulations adopted by the European Medicines Agency (EMA) and the American Food and Drug Administration (FDA). These regulations, known as the Good Scientific Practices (or GxP), are steeply transposed to a number of countries and their implementation is mandatory in the development of novel cell-based therapies, similar to traditional drug development in the pharmaceutical industry. However academic and not-for-profit institutions are the ones currently taking the challenge of adapting into this new scenario and opening the door to a new era in cell therapy.

Now, the increasing knowledge in the field, new technologies, the better understanding of diseases, and the possibility of "personalizing" cells and tissues, in a context of complex regulatory requirements, demands for professionals with specific competencies and deep understanding of this matter.

This is precisely the contribution of the present *Guide to Cell Therapy GxP*, addressed to any person working in the field, where the authors sew the key elements that compose the puzzle of this, I would say, revolutionary area of knowledge and patient care.

Joan Garcia
Divisi de Terpies Avanades/XCELIA, Banc de Sang i Teixits
Barcelona, Spain

Blanca Miranda
Biobank Andalousian Public Health Care System
Granada, Spain

References

[1] Mathe G, Jammet H, Pendic B, Schwarzenberg L, Duplan JF, Maupin B, et al. Transfusions and grafts of homologous bone marrow in humans after accidental high dosage irradiation. Rev Fr Etud Clin Biol March 1959;4(3):226–38.

[2] van Rood JJ, van Leeuwen A. Major and minor histocompatibility systems in man and their importance in bone marrow transplantation. Transpl Proc September 1976;8(3):429–36.

[3] Thomas ED, Buckner CD, Rudolph RH, Fefer A, Storb R, Neiman PE, et al. Allogeneic marrow grafting for hematologic malignancy using HL-A matched donor-recipient sibling pairs. Blood September 1971;38(3):267–87.

[4] Rosenberg SA, Restifo NP. Adoptive cell transfer as personalized immunotherapy for human cancer. Science April 3, 2015;348(6230):62–8.

[5] Orlando G, Soker S, Stratta RJ, Atala A. Will regenerative medicine replace transplantation? Cold Spring Harb Perspect Med August 1, 2013;3(8).

[6] Lysaght MJ, Crager J. Origins. Tissue Eng Part A July 2009;15(7):1449–50. 27.

[7] Lanphier E. State of the advanced therapies industry remarks. Paris: Alliance for Regenerative Medicine Summit; March 2015.

[8] Sibinga CS, Das PC. Quality assurance in blood banking and its clinical impact. Boston: Martinus Nijhoff Publishers; 1984.

Overview of the Development Program of a Cell-Based Medicine

1

Arnau Pla
CELLAB, Barcelona, Spain

Chapter Outline

1. Introduction

The goal of pharmaceutical product development is to establish the formulation composition and define its manufacturing process to consistently deliver a drug product. This drug product has to meet appropriate quality attributes required for its intended efficacy and safety profile. In addition to basic quality requirements, the commercial success of a drug product, and by extension its lifecycle, is determined by other key parameters such as patents, market, prices competence, regulatory changes, and others that must be carefully considered during early development stages. Pharmaceutical and biopharmaceutical industries have developed systematic approaches to fulfill these complex requirements. In contrast, the newborn cell therapy industry, closely linked to academia, should develop novel approaches to address this major challenge [1].

Although there are extensive resources and efforts devoted by many companies, to date, there are few cell therapy products licensed in Europe and in the United States (Table 1). This fact reflects the great complexity of developing such type of treatments. However, great hopes are invested in the emerging field of regenerative medicine and the use of cells as therapeutic agents. The term advanced therapy medicinal product (ATMP) covers the following medicinal products for human use (http://www.ema.europa.eu/ema/):

Table 1 **Approved Human Cell-Based Therapeutics**

Name; Description	Manufacture
In the United States:	
Provenge®; Autologous cellular immunotherapy	Dendreon Corporation
Laviv®; Autologous cultured fibroblasts	Fibrocell Technologies, Inc.
Carticel®; Autologous cultured chondrocytes	Genzyme BioSurgery
Gintuit®; Allogeneic cultured keratinocytes and fibroblasts in bovine collagen	Organogenesis, Inc.
Allocord®; HPC from cord blood	SSM Cardinal Glennon Children's Medical Center
Hemacord®; Allogeneic HPC from cord blood	New York Blood Center
Ducord®; HPC from cord blood	Duke University School of Medicine
HPC from cord blood	Clinimmune Labs, University of Colorado Cord Blood Bank
HPC from cord blood	LifeSouth Community Blood Centers, Inc.
In Europe:	
Chondrocelect®; Autologous cultured chondrocytes	TIGenix
MACI®; matrix-induced autologous chondrocyte implantation	Genzyme
Provenge®; Autologous cellular immunotherapy	Dendreon Corporation
Holoclar®; Autologous limbal stem cells	Chiesi Farmaceutici S.p.A.
In Canada and New Zealand [12]:	
Prochymal®; Adult human MSC	Osiris Therapeutics, Inc.
In Japan [12]:	
JACE®; Autologous cultured epidermis	Japan Tissue Engineering Company (J-TEC)
JACC®; Autologous cultured cartilage	Japan Tissue Engineering Company (J-TEC)
In Korea [12]:	
Hearticellgram-AMI®; Autologous bone marrow-derived MSC	Pharmicell
Cartistem®; MSC for the treatment of osteoarthritis	Medipost

With the exception of blood products, the rest include a substantial manipulation in their manufacture. Only approved human cell-based medicines were included. HPC = hematopoietic progenitor cells; MSC = mesenchymal stromal cells.

- **Gene-therapy medicines**: These contain genes that lead to a therapeutic effect. They work by inserting recombinant genes into cells, usually to treat a variety of diseases, including genetic disorders, cancer, or long-term diseases.
- **Somatic-cell therapy medicines**: These contain cells or tissues that have been manipulated to change their biological characteristics.

- **Tissue-engineered medicines**: These contain cells or tissues that have been modified so that they can be used to repair, regenerate, or replace tissue.
- **Combined advanced therapy medicines**: These are medicines that contain one or more medical devices as an integral part of the medicine.

Cell therapy-based medicinal products (CTMPs) are defined as medicinal products when there is more than minimal manipulation of the cellular component or where the intended use of the cells is different from their normal function in the body.

Much attention had been paid to the potential of novel stem cell- and tissue engineering-based therapies following a number of relevant scientific milestones and media news of potential new cures [2,3]. These therapies have become the focus of many biopharmaceutical developments, which face a number of major challenges in translating these scientific advances into Food and Drug Administration (FDA)/European Medicines Agency (EMA)-approved medical products. ATMPs, including cell therapy and tissue engineering products, are considered as medicines in the European Union [4].

ATMPs are at the forefront of scientific innovation in medicine; consequently, specific regulatory framework has been developed and implemented in Europe and in the United States. In this regard, the Regulation (EC) N° 1394/2007 on ATMPs was drafted and came into force in December 2008. The Regulation laid down specific rules concerning centralized authorization and pharmacovigilance of the ATMPs. This regulatory framework has a crucial influence in the development of such ATMPs. As a consequence, the teams involved in the ATMP development must take into account the FDA/EMA scientific and regulatory guidelines that provide a detailed description of the safety, efficacy, and quality issues for CTMPs [5,6]. As we can see in Box 1, cell therapy and tissue engineering products are very clearly defined by regulatory agencies.

Box 1 What are Cell Therapy and Tissue Engineering Products?

Cell therapies and tissue engineering products are considered as medicines when the following are true:

- Cell-based product:
  - Substantial manipulation of cells or not intended to be used for the same essential function(s)
  - Administered to human beings with a view to treating, preventing, or diagnosing a disease through the pharmacological, immunological, or metabolic action
- Tissue-engineered product:
  - Substantial manipulation of tissues or not intended to be used for the same essential function(s)
  - Engineered cells or tissues
  - Administered to human beings with a view to regenerating, repairing, or replacing a human tissue

---

2. Key Pharmaceutical Factors to Consider in Early Development Stages

Several technical obstacles must be overcome during the stages of product conception and design, before cell and tissue engineering therapies move out from basic research laboratories to clinical phases of investigation. There are many challenges associated with characterizing and quantifying the raw materials, cells, reagents, and processes for the therapies. From a regulatory point of view, ATMPs must be safe and effective and also be produced following high-quality manufacturing processes that allow for on-time delivery of viable products. CTMPs developers must be able to respond to the following:

- **Quality considerations**: Define a manufacturing process that is able to consistently produce sterile and pyrogen-free drug products with defined identity, purity, and potency, and produce a strong foundation in product development and cell characterization (as it relates to therapeutic efficacy of cell-based drugs).
- **Nonclinical considerations**: Adequately evaluate different aspects including proof of concept, dosage, biodistribution, persistence, and safety (i.e., tumorogenicity, immune rejection). In vitro or in vivo models may provide information, hence the importance of using biologically relevant preclinical models of safety and efficacy. Understand and control the mechanism of action and use of reliable potency assays. Establish the pharmacodynamics and pharmacokinetics and dose-finding studies.
- **Bioprocess considerations**: What is required to create scalable and robust manufacturing processes [7] are considered. What are the basic unit operations: cell expansion, centrifugations. Economic cost of the cell production process should be carefully considered to avoid unrealistic approaches.
- **Stability**: Due to the living nature of ATMP, stability is one of the most challenging aspects. This aspect should be studied from the beginning. Long and noncomplex conditions of storage are desirable. The formulation and packaging strongly influence stability characteristics. Once the product has been totally developed, long-term stability programs should be performed.
- **Product characterization**: Develop potency tests [8]. To establish the activity of the product, adequate tests must be developed. These tests should mimic, as much as possible, the desired biological function to be developed by the ATMP inside the organism. After the development of the test, limitations such as economic or time restrictions due to the limited stability of the final product have to be evaluated.
- **Packaging/administration considerations**: A good product with complex packaging or administration system can be rejected by the clinicians due to intrinsic complexity of the surgery. Simple and secure packaging forms should be considered. For injectable drug products, prefilled injections are preferable.

The lack of drug development know-how and experience can lead to the definition of products that may not be feasible for clinical evaluation or future commercialization. A poor understanding of the drug product characteristics and ineffective process development can impair critical product characteristics and lay the foundation for clinical failure. In this regard, attention should be paid in the early stages of development to clearly define the characteristics with special consideration of the regulatory and clinical practitioners needs in mind (Table 2).

Table 2 Analysis of Potential for Errors and Recommendations in the Development of Cell-Based Therapeutics

Potential for Errors	Recommendation
• Lack of expertise in regulatory matters • Adding unnecessary complexity in all phases of the product development pipeline • Deadlines too optimistic • Overacting due to lack of experience and risk aversion • Underestimation of the investment (financial, time, and personnel) required	• Establish contact with regulatory authorities early in the development and make use of scientific advice • Use CROs and support services
• Performing studies without proper documentation	• Early implementation of GxP quality assurance systems based on existing certifications (such as ISO9001, JACIE, and FACT-NetCord)
• Not establishing clearance for product commercialization	• Patent policy to ensure FTO
• Development of cell products that require complicated clinical protocols or are difficult to deliver	• Participation of surgeons in preclinical studies, so their inputs are taken into account early in the product development process

CBMP = cell-based medicinal product; CRO = contract research organization; FTO = freedom to operate.
Adapted from Vives et al. [13].

3. TPP: Beginning with the End in Mind

The use of a systematic approach such as the target product profile (TPP) can facilitate the right regulatory and business approaches, determine market segments, and create an understanding of the competition and associated pricing strategies. A TPP is a key strategic document that summarizes features of an intended commercial therapeutic product. The definition of a TPP according to the target-disease health requirements and user needs should drive the design of other fundamental aspects as the primary container-closure system, the stability requirements, or the logistic approach. This later point is critical because of the special features of cell therapy and tissue engineering products. This reflection and the effort to understand the problem result in improved provider/patient convenience and regulatory compliance, as well as product differentiation. Further, the use of TPP will enable achieving an integrated approach to product and process development that can lead to clinical and commercial success [9].

The TPP is an organized list, developed and agreed upon from multiple stakeholder perspectives, which prioritizes the key features and attributes of the intended end product (the marketed drug). The TPP is a dynamic, evolving, written document, and is a focal point of reference for the project. It covers at least the following areas:

• Strategic context
• Medical and commercial requirements and priorities

Box 2 TPP basic features for cell-based medicinal products

Features of a TPP

- Indications and use
- Productive process
- Dosage and administration
- Product characteristics
- Contraindications
- Conservation and stability
- Industrial protection
- Regulatory strategy

- Patient requirements
- Technical (research) and biological features
- Technical (development) requirements and feasibility
- Desired versus minimally acceptable features

The TPP is a valuable tool to organize all relevant information from multiple perspectives (i.e., medical, market, production, regulatory). This "living" document can be used to help us get a right relationship between development and attributes of a desired ATM drug (see Box 2).

4. Stages of Drug Development

The length of the cell therapy pharmaceutical product development from discovery to market might take several years with extensive financial and scientific risks (Figure 1). The process is long and uncertain, and in addition, as soon as market arrival, the product developed must also compete with existing treatments. Taking into account this complex scenario, multiple aspects complementary to product conception, design, and development should be considered to increase the possibility of market arrival and commercial success.

Any drug development process must proceed through several stages, as detailed next:

- **Discovery**: Once the TPP is defined, the development process begins with the performance of pending aspects related to basic research or literature review, in which the regenerative basis of diseases are targeted, studied, and then therapies are proposed. A key milestone of this development stage is the understanding and validation of the proof-of-principle. This involves obtaining strong in vitro and animal data. When the therapeutic approach has been developed and proof-of-principle validated, intellectual property issues must be addressed to ensure freedom to operate in the area of interest. This is a critical step to attract or maintain investors and to prevent ideas from being appropriated by a third party. Without a patent, investors will not feel their high-risk investment is safe from others simply copying the process or product once it is approved. Patents are most commonly used for cell and tissue therapy and can protect an innovation for 20 years.

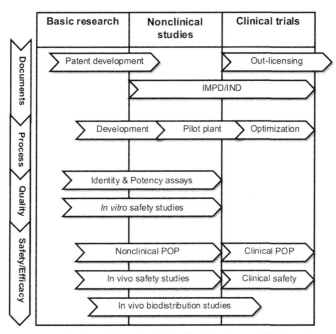

Figure 1 Development pipeline. Flowchart depicting the phases of the development of a cell-based therapeutic up to the Phase I/IIa clinical trial. IMPD = investigational medicinal product dossier; IND = investigational new drug; POP = proof-of-principle.
Adapted from Vives et al. [13].

- **Preclinical studies**: The objective of preclinical studies is to test the safety of the products and to gain insight into the working mechanisms of the cell therapies. These studies should be done before the submission of the Common Technical Document (CTD) to EMA, and complementary data can be obtained during clinical trials. Nonclinical evidence on the proof-of-principle and safety of the stem cell-based product in a relevant animal model is expected.
- **Clinical trials**: Clinical trials should be designed to demonstrate safety and efficacy as well as evidence to substantiate the mode of action identified during the clinical trial [10]. Briefly, once the EMA approves a drug for clinical trials, Phase I is performed to check the safety and concept proof of the drug. If a drug passes Phase I testing (or Phase I/IIa, for most ATMP developments) and is considered safe for humans, the drug product is then tested on patients in Phase II clinical trials. Phase II studies involve a higher number of patients divided into subgroups to test different doses. The data obtained in Phase II clinical trials are used for the primary purpose of designing optimal Phase III clinical trials. Phase III trials will be used as a basis for obtaining statistically significant data to prove the effectiveness and to warrant the safety of the drug. This development stage is known as the "pivotal trials" because the data generated at this stage are used to obtain regulatory approval.
- **Approval and post-marketing testing**: Finally, if a drug passes Phase III clinical studies and the new drug application (NDA) is approved, the drug can be marketed for sale to the public.

At every stage of drug development (discovery, nonclinical, clinical, and commercialization), specific quality and regulatory frameworks must be applied: good laboratory practices, good manufacturing practices, and good clinical practices. Therefore, companies conducting ATMP developments must have a quality system (QS) with the appropriate Good Scientific Practice regulations (GxP) in place, which are inspected and certified by regulatory authorities. The QS must guarantee various aspects:

- Quality management
- Quality assurance
- Risk management tools
- Continuous improvement

Implementing and maintaining efficient QS can represent a complex task and involves up-front costs. In this regard, quality cannot be an afterthought. Implementing an effective quality system should be in place at the earliest stages of ATMP research and development.

5. Considering Stakeholders

To ensure product development and lifecycle success, many different resources should be recruited and mobilized—not only economic but also scientific, regulatory, techno-logic, legal, and others. Once a potential cell- or tissue-based therapy has been identified and defined, all stakeholders in the product development process must be identified, considered, and so far as they are necessary involved. In this regard, the nature of the company (public/government-sponsored or private capital founded) should be considered. Usually, privately funded companies have very different return reimbursement rates than public/government-sponsored ones. Moreover, due to the nature of public/govern-ment-sponsored institutions, these can be more committed to nonprofit medical problems.

Stakeholders can be broadly divided into five main groups:

- **Scientific/technologic**: Technologic partners can provide key solutions to bioprocess require-ments, product characterization, or other scientific needs. For example, Good Laboratory Practice (GLP)/ Good Manufacturing Practice (GMP) laboratories capable of performing under demand, preclinical studies, or batch production and release can be so interesting as to avoid extensive inversions. Externalization of services is a good strategy to diminish the global inversion value.
- **Regulatory**: As soon as the preclinical and clinical plans are defined, regulatory authorities such as EMA/FDA should be involved in product development.
- **Investors**: The long time periods required to reach market require the collaboration with stable, well-planned, and long-term financial partners. Growing investment needs will be required during the development. In this regard, different investor profiles can be considered depending on the development stage: business angels, venture capital funds, or corporate funds. A well-defined and pre-established business strategy is an essential factor to guarantee the arrival to the market.
- **Clinicians**: Patient recruitment, procurement of tissues or cells, and intensive collaboration on defining product characteristics, therapeutic approach, and packaging should be discussed with clinicians.
- **Customers**: In the era of easy access to information, the paperwork for the final user of the medicines must be considered. Now, not only the clinician is the final user, but also the final user experience can determine the success or failure of a cell therapy treatment.

6.   Product Lifecycle and Portfolio Management

One of the best opportunities for increasing the growth of income returns in pharmaceutical companies is through the launching of new products that open new markets or beating the competence on existing markets through innovation, which leads to competitive advantages. To do this, organizations need to understand how to create value for their customers/clinicians/patients. The process of value creation should involve the organization's own strengths and avoid weaknesses contributing to product development failure. But even the most successful company with the most successful products are subjected to product lifecycles. Long-established products or treatments eventually become less popular; whereas, the demand for new, more modern, and effective treatments usually increases. Companies who fail to launch new products have little chance of survival. Business dynamics generated by the product's lifecycles oblige the companies to maintain a portfolio of potential new products to stay competitive by early arrival to market. The importance of being the first lies in the benefits derived, such as longer sales life of the product, increased margins, or increased product loyalty. Speed to market is a key variable and can determine the success or failure of a new product or company. Because of this, the prioritization of new product/innovation projects is a critical management task (Figure 2).

Usually, small and large companies develop several projects simultaneously, consuming great amounts of human, material, and monetary resources. All these projects need to be balanced with respect to a range of parameters including regulatory, patents, market, risk, money, and competence, among others. From this point of view, there is a strong rationale for linking portfolio management closely with organization strategy.

In a global context of elevated business complexity, the ability of a company to manage its portfolio of developments has become a key competence. The implementation of a portfolio's management in the earlier phases will likely increase the success

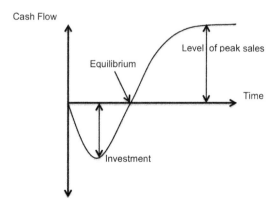

Figure 2 Graphical representation showing the different stages of the lifecycle of the development of pharmaceutical drugs. Next to investment, the benefits grow until the plateau phase is reached. Before this point is reached, new products should be developed and launched to the market.

rate of a company. Simultaneously, a critical analysis of what went wrong when negative results are found in clinical trials may help to design better strategies [11].

7. Performance Management and the Check Point Value

In the complex scenario of ATMPs development, success does not only rely on having relevant stakeholders, promising products, brilliant scientists, abundance of resources, or a strong intellectual property, but it also depends on proper management of the development process itself. During the development process, the coexistence of several complex research activities taking place at the same time, such as preclinical, production, regulatory issues, etc., introduces a real risk of losing concentration on the target product and may incur major problems. This is especially true in academic environments, where scientists are prone to further investigate observations that may not be relevant to the ATMP development itself.

Because of these reasons, organizations developing extensive R & D activities could be greatly benefited from implementing and systematizing accurate performance management (PM) tools. PM should be in line with the company mission, vision and strategy, and goals, and it serves as powerful tool for aligning all development activities toward the desired direction.

Although PM should be planned, implemented, and assured by the management team, all the members of the R & D team should be involved early in the performance appraisal. Explaining the objectives, fixing incentive strategies, and establishing the time periods and the deadlines can help to assure a good enrollment of the team and serve to avoid later misunderstandings.

PM has several key points:

1. **Fixing the objectives**: These objectives must be concise, measurable, and more importantly, achievable. Further, the goals have to be communicated to all in the organization.
2. **Evaluation of the process and incentives**: Performance has to be measured and evaluated at several time points to maximize the possibilities of success and detect deviations or critical delays. The use of an incentives policy may serve as a value creator.
3. **Learning and improvement**: All the expertise gained during the process can be applied to overcome future difficulties through learning from our previous mistakes.

In addition to the previous points, corrective actions can be designed, and organizations can use several indicators for PM, either financial or nonfinancial (such as operational or process indicators). Among them, the use of check point values (CPVs) can be a very useful tool that gives the chance to adjust the priorities or resolve issues. CPV can track several aspects and provide R & D managers with complex and complete information whether development is running the right way.

An example of a CVP addressed to evaluate some of the relevant aspects of the development of any ATMP is shown in Table 3.

Table 3 **Checkpoint Value for ATMP Development**

Objective	Indicator	Mitigation Actions
Preclinical toxicology	Final report demonstrating safety of the drug product	New study on relevant animal species
Intellectual property	Achievement of PCT	New aspects increasing the robustness of the patent
Process	Process capable of producing the final dose into a reproducible manner	Introducing new technologies to accelerate the production time
Proof of concept	Obtaining relevant data about the mechanism of action of the drug product	Determine other potential mechanisms
Regulatory	Obtain GLP certification of the lab to assure the quality of preclinical data	Establish collaboration with accredited GLP labs to perform most relevant works
	GMP certification of the production facility	Correct major deviations and send to regulatory authorities
Quality control	Potency test able to correlate in vivo and in vitro activities	Establish new relevant cytometry markers
Phase I clinical results	Obtain relevant data concerning product safety properties	New nonclinical study on relevant animal species to determine toxicity

PCT = Patent Cooperation Treaty; GLP = Good Laboratory Practice; GMP = Good Manufacturing Practice.

The use of managerial tools to fix, measure, and establish the grade of achievement of objectives is a widely extended and validated methodology, and it can help to assure the correct development of the process. In particular, its use is recommended at several time points of the development process.

8. Conclusions

ATMPs, defined as gene-therapy medicinal products, somatic-cell therapy medicinal products, and tissue-engineered products, as well as combined advanced therapy medicines, constitute innovative therapeutics that are being investigated as novel treatments for several medical situations. Despite major economic and human efforts dedicated to the development of such products, only a few cell-based medicines have reached the marketing authorization in the European Union or United States. Academia, charities, and small companies are the main protagonists of these laudable efforts trying to solve important health problems. These institutions usually have limited access to funding and less extensive experience on developing pharmaceutical products than big pharma companies. Both of them are indispensable to achieve a marketing authorization or use under the hospital exemption clause by establishing collaborations at mid/late product development stages. These circumstances can act as limiting factors of the capacity to develop successful strategies and products. The understanding of the development process of a cell-based medicine may have a direct impact on the schedule and investment required for successfully bringing candidate therapies from the bench to the clinics. Moreover, it will help the developer to detect whether the candidate medicine may be of interest to investors or will necessarily have to be government-sponsored to reach the clinics.

References

[1] Maciulaitis R, D'Apote L, Buchanan A, Pioppo L, Schneider CK. Clinical development of advanced therapy medicinal products in Europe: evidence that regulators must be proactive. Mol Ther March 2012;20(3):479–82.

[2] Alper J. Geron gets green light for human trial of ES cell-derived product. Nat Biotechnol March 2009;27(3):213–4.

[3] Cyranoski D. Stem cells cruise to clinic. Nature February 28, 2013;494(7438):413.

[4] Committee for Advanced Therapies, Secretariat CATS, Schneider CK, Salmikangas P, Jilma B, Flamion B, et al. Challenges with advanced therapy medicinal products and how to meet them. Nat Rev Drug Discov March 2010;9(3):195–201.

[5] Bailey AM, Mendicino M, Au P. An FDA perspective on preclinical development of cell-based regenerative medicine products. Nat Biotechnol August 2014;32(8):721–3.

[6] Mendicino M, Bailey AM, Wonnacott K, Puri RK, Bauer SR. MSC-based product characterization for clinical trials: an FDA perspective. Cell Stem Cell February 6, 2014; 14(2):141–5.

[7] Sart S, Schneider YJ, Li Y, Agathos SN. Stem cell bioprocess engineering towards cGMP production and clinical applications. Cytotechnology October 2014;66(5):709–22.

[8] Bravery CA, Carmen J, Fong T, Oprea W, Hoogendoorn KH, Woda J, et al. Potency assay development for cellular therapy products: an ISCT review of the requirements and experiences in the industry. Cytotherapy 2013;15(1):9–19.e9.

[9] Curry S, Brown R. The target product profile as a planning tool in drug discovery research. 2003.

[10] Lee MH, Arcidiacono JA, Bilek AM, Wille JJ, Hamill CA, Wonnacott KM, et al. Considerations for tissue-engineered and regenerative medicine product development prior to clinical trials in the United States. Tissue Eng Part B Rev February 2010;16(1):41–54.

[11] Galipeau J. The mesenchymal stromal cells dilemma–does a negative phase III trial of random donor mesenchymal stromal cells in steroid-resistant graft-versus-host disease represent a death knell or a bump in the road? Cytotherapy January 2013;15(1):2–8.

[12] Mason C, Mason J, Culme-Seymour EJ, Bonfiglio GA, Reeve BC. Cell therapy companies make strong progress from October 2012 to March 2013 amid mixed stock market sentiment. Cell Stem Cell June 6, 2013;12(6):644–7.

[13] Vives J, Oliver-Vila I, Pla A. Quality compliance in the shift from cell transplantation to cell therapy in non-pharma environments. Cytotherapy March 10, 2015. pii: S1465-3249(15)00049-3. http://dx.doi.org/10.1016/j.jcyt.2015.02.002. [Epub ahead of print].

Glossary

Freedom to operate Evaluation of whether you infringe the patent, design, or trademark rights of another entity.

Target product profile Document that summarizes the features of an intended commercial therapeutic product.

Orphan drug In cell-based therapies, an ATMP that has been developed specifically to treat a rare medical condition.

Patent Cooperation Treaty International agreement for filing patent applications, having effect in more than 100 countries.

List of Acronyms and Abbreviations

ATMP Advanced therapy medicinal product
CRO Contract research organization
CTD Common Technical Document
CTMP Cell therapy medicinal product
CPV Check point value
GCP Good Clinical Practice
GLP Good Laboratory Practice
GMP Good Manufacturing Practice
EMA European Medicines Agency
FDA Food and Drug Administration
FTO Freedom to operate
R&D Research and development
PCT Patent Cooperation Treaty
PM Performance management

European Regulatory Framework for the Development of Cell-Based Medicines

2

Natividad Cuende[1], Ander Izeta[2]
[1]Andalusian Initiative for Advanced Therapies. Servicio Andaluz de Salud, Consejería de Salud, Junta de Andalucía. Sevilla. Spain; [2]Tissue Engineering Laboratory, Bioengineering Area, Instituto Biodonostia, Hospital Universitario Donostia, San Sebastian, Spain

Chapter Outline

1. Introduction

For any cell biologist specialized in translational research on a particular cell-based therapy, achievement of positive endpoint results in a proof-of-concept animal model may be one of the most exciting moments in his/her lifetime, due to the potential harnessing of application to human beings after a long and sometimes tortuous R&D program. However, that same moment could also mark the commencement of one of the most challenging and sometimes frustrating periods of their professional careers, if they decide to go a step further translating those findings into the clinical setting.

The main reason for this apparent contradiction is the complexity of the legal framework regulating cell-based products, which are technically complex per se. Most cell-based products are considered as medicines in Europe, representing a novel class named advanced therapy medicinal products (ATMPs) that comprise the following categories:

- Somatic cell therapy medicinal products (SCTMPs)
- Gene therapy medicinal products (GTMPs)
- Tissue engineered products (TEPs)
- Combined advanced therapy medicinal products

2. What Cell-Based Products are Considered as Medicinal Products? The Legal Definitions and Main Regulations Applying to Cell-Based Products

The current legal definitions of cell and GTMPs are found in Directive 2001/83/EC [1] as amended by Commission Directive 2009/120/EC [2], and the definitions of TEPs and combined ATMPs in Regulation (EC) No 1394/2007 [3] (Figure 1).

2.1 Gene Therapy Medicinal Products (GTMPs)

Although not all GTMPs involve cells, in the case of ex vivo gene therapy, cells play an essential role. Usually, we identify the concept of gene therapy simply as the insertion, alteration, or removal of genes within individual cells and biological tissues to treat a disease. Nevertheless, in gene therapy, frequently, but not necessarily

Figure 1 Regulations that define the different ATMPs in Europe.

always, a recombinant vector, which can be viral or nonviral, with the therapeutic gene is used for gene delivery to specified cells and tissues. Two different strategies are used for this gene delivery, that is, ex vivo and in vivo. In the ex vivo approach, the cells may be cultured and used for gene transfer, so that these transduced cells are then introduced in a target tissue. Alternatively, in the in vivo approach, the gene may be delivered through a vector directly into the target cell or tissue.

The most common form of gene therapy involves the insertion of functional genes into an unspecified genomic location to replace a mutated gene, but other forms involve directly correcting the mutation or modifying normal genes, for example, to make a patient resistant to a viral infection or to increase the production of a functional protein.

From the regulatory point of view, the definition of a GTMP is more precise regarding the objectives and effects, and it excludes vaccines from its scope. The definition is found in the Commission Directive 2009/120/EC [2]:

"Gene therapy medicinal product means a biological medicinal product which has the following characteristics:

1. It contains an active substance which contains or consists of a recombinant nucleic acid used in or administered to human beings with a view to regulating, repairing, replacing, adding, or deleting a genetic sequence.
2. Its therapeutic, prophylactic, or diagnostic effect relates directly to the recombinant nucleic acid sequence it contains, or to the product of genetic expression of this sequence.

Gene therapy medicinal products shall not include vaccines against infectious diseases."

2.2 Somatic Cell Therapy Medicinal Products (SCTMPs)

One of the simplest ways to define cell therapy can be the use of cells to treat a disease. This includes any type of cell, irrespective of its source (human-autologous or allogeneic and animal), the degree of differentiation (committed cells, progenitors, or stem cells), or their origin (embryo, fetus, newborn, or adult individuals).

A further issue concerns the concept of somatic cell therapy products within the scope of medicinal products. This definition is also found in the Commission Directive 2009/120/EC [2]:

"Somatic cell therapy medicinal product means a biological medicinal product which has the following characteristics:

1. Contains or consists of cells or tissues that have been subject to substantial manipulation so that biological characteristics, physiological functions, or structural properties relevant for the intended clinical use have been altered, or of cells or tissues that are not intended to be used for the same essential function(s) in the recipient and the donor.
2. It is presented as having properties for, or is used in or administered to human beings with a view to treating, preventing, or diagnosing a disease through the pharmacological, immunological, or metabolic action of its cells or tissues."

For the purposes of point (1), the manipulations listed in Box 1—as detailed in Annex I to Regulation (EC) No 1394/2007 [3]—shall not be considered as substantial manipulations.

Box 1 Nonsubstantial Manipulations

- Cutting
- Grinding
- Shaping
- Centrifugation
- Soaking in antibiotic or antimicrobial solutions
- Sterilization
- Irradiation
- Cell separation, concentration, or purification,
- Filtering
- Lyophilization
- Freezing
- Cryopreservation
- Vitrification

2.3 Tissue Engineered Products (TEPs)

The first idea that springs to mind when we are speaking about TEPs is a scaffold, more or less complex, biological or not, in combination with cells. Although this may indeed be typical of a tissue engineered product, a scaffold does not necessarily need to be present for a product to be included in this category of ATMPs.

The definition is set out in the Regulation (EC) No 1394/2007 [3] as follows:

"Tissue engineered product means a product that:

- contains or consists of engineered cells or tissues, and
- is presented as having properties for, or is used in or administered to human beings with a view to regenerating, repairing, or replacing a human tissue.

A tissue engineered product may contain cells or tissues of human or animal origin, or both. The cells or tissues may be viable or nonviable. It may also contain additional substances, such as cellular products, bio-molecules, biomaterials, chemical substances, scaffolds, or matrices.

Products containing or consisting exclusively of nonviable human or animal cells and/or tissues, which do not contain any viable cells or tissues and which do not act principally by pharmacological, immunological, or metabolic action, shall be excluded from this definition."

Therefore, the presence of a matrix or scaffold is not necessary for a product to be considered a tissue engineered product. The key is the presence of engineered cells or tissues and the objective of its administration. When considering the concept of engineered cells or tissues, the Regulation states that

"Cells or tissues shall be considered 'engineered' if they fulfill at least one of the following conditions:

- The cells or tissues have been subject to substantial manipulation, so that biological characteristics, physiological functions, or structural properties relevant for the intended regeneration, repair, or replacement are achieved.
- The cells or tissues are not intended to be used for the same essential function or functions in the recipient as in the donor."

If we review the aforementioned definition of SCTMPs, we will see that the composition of both SCTMPs and TEPs may be identical. Consequently, the difference rests in the second condition related to the objective of its administration. Their mode of action is different: whereas TEPs are administered with a view to regenerating, repairing, or replacing a human tissue, in the case of SCTMPs, they are administered with a view to treating, preventing, or diagnosing a disease through the pharmacological, immunological, or metabolic action of its cells or tissues.

One example of a product that can be considered as SCTMP or TEP could be substantially manipulated mesenchymal stem cells. When they are used for immunomodulation to treat an autoimmune disease, they might be classified as SCTMP; whereas, they might be classified as TEP when used to repair a bone fracture.

Hence, a TEP might be simply defined as a product containing cells or tissues that have been engineered so that they can be used to repair, regenerate, or replace tissue.

2.4 Combined Advanced Therapy Medicinal Products

The last category of ATMPs comprises the combined ATMP. Regulation (EC) No 1394/2007 [3] defines a combined ATMP as that which

"fulfills the following conditions:

- It must incorporate, as an integral part of the product, one or more medical devices or one or more active implantable medical devices, and
- Its cellular or tissue part must contain viable cells or tissues, or
- Its cellular or tissue part containing nonviable cells or tissues must be liable to act upon the human body with action that can be considered as primary to that of the devices referred to."

As in the previous cases, the Regulation also considers that

"where a product contains viable cells or tissues, the pharmacological, immunological, or metabolic action of those cells or tissues shall be considered as the principal mode of action of the product."

2.5 Limits between ATMPs Categories

Taking these possibilities into account, some products could fall into different categories of ATMPs, but the Regulation (EC) No 1394/2007 [3] clarifies this:

"A product which may fall within the definition of a somatic cell therapy medicinal product or a tissue engineered product, and a gene therapy medicinal product, shall be considered as a gene therapy medicinal product."

And in cases in which the difference between an SCTMP and a TEP is not clear, the Regulation establishes that

"A product which may fall within the definition of a tissue engineered product and within the definition of a somatic cell therapy medicinal product shall be considered as a tissue engineered product."

Regulation (EC) No 1394/2007 [3] establishes that any applicant developing a product based on genes, cells, or tissues may request a scientific recommendation of the European Medicines Agency (EMA) to determine whether the referred product falls within the definition of an ATMP. The EMA shall deliver this recommendation after consultation with the Commission and within 60 days after receipt of the request. Those recommendations are available, after deletion of all information of commercial confidential nature, on the EMA website. The Regulation also establishes the creation of a Committee for Advanced Therapies (CAT) within the EMA. The CAT can provide advice on whether a product falls within the definition of an ATMP. To request ATMP classification, a Pre-submission request form and Briefing Information (including background information on scientific, legal, regulatory, and medical aspects) have to be completed and sent to AdvancedTherapies@ema.europa.eu.

In Figure 2, we can see a decision tree (available at the Paul-Ehrlich-Institut Website [4]) that may be helpful in order to know if a product falls within the ATMP group and to classify it according to the different categories.

2.6 Borderline Products

The legal definitions of the different types of ATMPs incorporate concepts not easy to apply in specific cases. Sometimes, it is not easy to decide what category corresponds with a specific product, and it is even more difficult to decide if a cell-based product is indeed a medicinal product or not. To provide guidance on the ATMP classification procedure as well as on the interpretation of the legal concepts, the EMA/CAT published a reflection paper on classification of ATMPs that has been updated to reflect the current thinking of the CAT on what medicines can or cannot be classified as ATMPs. The updated reflection paper was adopted by CAT in May 2015 after public consultation [5] and recognizes the difficulty as long as it discusses some borderline cases and areas where scientific knowledge is limited or evolving rapidly.

The key points regarding the limits to determine whether a cell-based product is or is not a medicinal product are based on the following:

- The type of manipulation performed on the cells (whether they have been subjected to substantial manipulation so that biological characteristics, physiological functions, or structural properties relevant for the intended clinical use have been altered)
- The intended use—for the same or different essential function or functions—in the recipient and the donor, irrespective of whether the donor and the recipient are the same person

Regarding what is considered substantial manipulation, the Regulation only provides a nonexhaustive list of the manipulations not considered as substantial (Box 1). The culturing of cells—one of the most common manipulations performed in the use of cells as a therapy—is generally considered substantial manipulation. In fact, according to the reflection paper, the CAT considers substantial manipulation cell culturing leading to expansion. Therefore, when a cell-based product is subjected to substantial manipulation, even if we are intending to use it for the same function in the recipient as in the donor, then we are dealing with a medicinal product.

Bearing in mind the conditions with which a cell product must comply to be considered as a medicinal product, we can find some examples of cell therapies

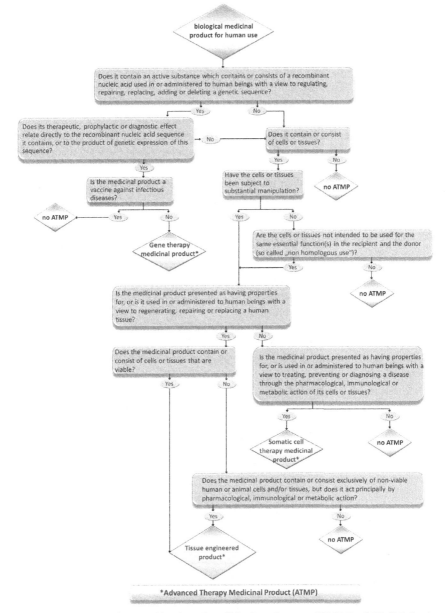

Figure 2 Decision tree for classification of medicinal products as ATMP. Paul-Ehrlich-Institut. Available in: http://www.pei.de/SharedDocs/Downloads/EN/pu/innovation-office/decision-tree-at-mp.pdf?__blob=publicationFile&v=1 [accessed 06.03.15].

that are considered to be transplants rather than medicinal products. This is the case of bone marrow transplantation in which the bone marrow progenitors are nonsubstantially manipulated and the intended use is to replace the hematopoie-sis in the recipient. Since among bone marrow progenitors there is a significant number of hematopoietic progenitors, we can consider hematopoiesis to be one of the essential functions of these bone marrow progenitors in the donor. Another example of a cell therapy not considered as a medicinal product is the transplantation of pancreatic islets when these are not cultured before being transplanted. In this case, they are only purified, therefore nonsubstantially manipulated, and their intended use is the production of insulin, which represents the same essential function in the recipient as in the donor.

Nevertheless, there are other cases in which cellular products have not been substantially manipulated, and where it is not easy to determine if their function is substantial or not [6]. The concept paper reflects the interpretation made by CAT incorporating the concept of nonhomologous use of cells or tissues (not considered in the European legislation) as equivalent of their use for a different essential function leading to classifications subject of debate [7].

2.7 Main Regulations Applying to Cell-Based Products

It has been mentioned that not all cell-based products fall within the scope of the definition of SCTMPs, ex vivo GTMPs, TEPs, or combined ATMPs. Some cell-based products are considered as transfusion or transplants, and their development is regulated under a different legal framework.

Later in this chapter, the general legal framework will be explained that applies to ATMP development; therefore, here we will only underline the similarities and differences in the development of a cell-based product considered a medicinal product or not (Figure 3).

While blood cells intended for transfusion, mainly regulated through Directive 2002/98/EC [8] among others [9,10], do not share any regulation with the other two types of cell-based products, ATMP and cell transplantation share some regulation, and sometimes, the boundaries between their definitions are blurred, making classification difficult [6], as mentioned above.

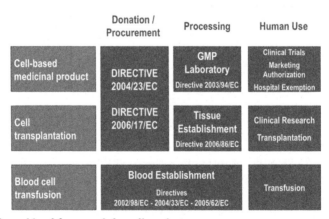

Figure 3 General legal framework for cell products.

It is important to take into account that, irrespective of whether they are considered as medicinal products or transplants, all cell-based products have to comply with the standards of quality and safety regarding donation, procurement, and testing. These standards are established in the Directive 2004/23/EC [11] on setting standards of quality and safety for the donation, procurement, testing, processing, preservation, storage, and distribution of human tissues and cells and in the Commission Directive 2006/17/EC [12] implementing Directive 2004/23/EC regarding certain technical requirements for the donation, procurement, and testing of human tissues and cells.

Concerning the requirements for processing these types of products, in the case of a medicinal product, Good Manufacturing Practice (GMP) compliant facilities are required, and the principles of GMP should be followed [13]. This will be examined below at length. In the case of cell or tissue transplants, a tissue establishment is required, and it is also necessary to comply with the requirements set out in Commission Directive 2006/86/EC [14] implementing Directive 2004/23/EC [11] regarding traceability requirements, notification of serious adverse reactions and events, and certain technical requirements for the coding, processing, preservation, storage, and distribution of human tissues and cells.

Finally, the clinical use is regulated differently according to the phase of development (experimental or not) and the nature of the cell product. As we will see, in the case of medicinal products, their clinical use must follow the clinical trial regulation—when they are still considered as investigational medicinal products— or the product has to be granted marketing authorization by the EMA. There could also be another possibility for ATMPs not industrially prepared under the denominated "hospital exemption" scheme. The clinical use of cell or tissue transplants— once their safety and efficacy have been demonstrated in clinical research—is also regulated by Commission Directive 2006/86/EC [14]. The authorization pathway is completely different, not involving Medicines Agencies.

3. An Introduction to Cell-Based Medicine Development: Roadmap

Cell-based medicines are a particularly novel class of medicines and possibly constitute one of the most complex tasks that may be approached by clinical researchers when exploring new therapeutic applications. In Europe, ATMPs, including cell therapy, gene therapy, and TEPs, represent a field with a constantly evolving regulatory landscape that scientists and regulators alike find difficult to navigate. Stem cell scientists should, therefore, be aware of the intricacies of GMP implementation before initiating full-fledged translational programs, and they should also have at their disposal well-trained technologists to develop ATMPs in different laboratories and institutions—be it hospitals, academia, or industry—within Europe [15].

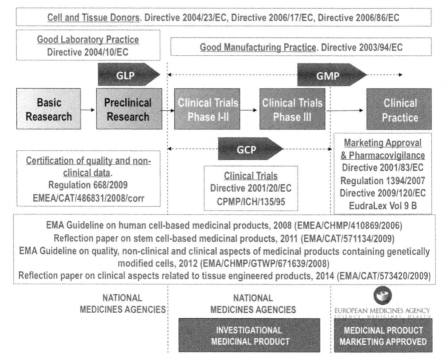

Figure 4 Relevant European rules and guidance for the ATMP development.

In as far as ATMPs are considered, a particular category of biological medicines, their development must not only fulfill the requirements for a medicinal product but also adhere to some very specific rules, and moreover, a set of EMA guidelines, concept papers, and reflection papers should be followed. Nevertheless, ATMPs are highly heterogenous, and regulatory authorities will always apply their rulings on a case-by-case basis.

Summarized in Figure 4 are the most important European rules and guidelines that must be taken into account to develop a stem cell medicinal product or other ATMPs. In the case of cell-based medicinal products, the standards of quality and safety for donation, procurement, and testing of human tissues and cell donors must also be followed.

At this point, it is important to remind the reader that this chapter will serve as an introduction for subsequent chapters that will approach specific issues in a more comprehensive manner. Therefore, the objective of this chapter is to provide a general picture to facilitate the integration of latter sections into the general roadmap for the development of ATMP. A helpful tool to identify and download the most relevant regulation and guidelines related to the ATMP development is shown in Box 2 at the end of this section.

Box 2 Additional Information on ATMP Regulation

The web page of the Andalusian Initiative for Advanced Therapies is a good resource to consult and download the main European regulations (Directives and Regulations) that regulate basic, preclinical, and clinical research with these kinds of products, their quality, and manufacturing aspects, as well as their marketing or clinic use. There is also a selection of EMA and ICH guidelines and other documents of interest related to these issues including FDA regulations and Pharmacopeias from Europe and the United States [82].

3.1 Aspects to Consider When Designing Proof-of-Concept Experiments in Animal Models

Let us imagine we have a reasonably characterized cellular product that we intend to put into human beings to treat a condition for which there is a rationale at least for a purported therapeutic effect. Human cell-based medicinal products are highly heterogenous, and regulatory authorities will always apply their rulings on a case by-case basis. When reviewing an investigational medicinal product dossier (IMPD) application, national authorities will usually require the identification of risk factors inherent in the nature of the ATMP in question and associated with its quality, safety, and efficacy [16].

As a rule of thumb, regulatory requirements will usually be less stringent for early clinical trials and will increase sharply as we approach marketing authorization. However, this is not always the case, for example, safety data for Phase I studies with stem cells may be very stringent. In any case, it may be counterproductive to rush into clinical trial testing before solid product characterization and nonclinical data are available, since, at the end of the day, the evaluation process will be further lengthened.

There are some principles that may be generally applied to all cell-based products. A good starting point for a newcomer would be the relevant chapters of the European Pharmacopoeia (Ph. Eur.) [17], EMA guidelines on cell therapy and tissue engineering [18–21], and those on gene therapy [22]. In addition, in the case of gene therapy product development, the guideline on quality, nonquality, and clinical aspects of medicinal products containing genetically modified cells should be consulted [23]. Of note, researchers in the advanced therapies area should be aware of the fast pace of regulatory changes that affect product development, with new guidelines arising every few months.

Before administration into humans, both biodistribution and toxicity of the investigational medicinal product must be tested in a relevant animal model according to Good Laboratory Practice (GLP) [24–26]. These usually involve subcontracting a contract research organization (CRO) specialized in basic pharmacology, toxicology, or safety studies so that the reports issued will comply with regulations. However, this is not always the case: contracts with CROs may be prohibitive for many laboratories, and national regulatory agencies will sometimes accept more basic laboratory studies provided "GLP-like" conditions have been followed according to the relevant guidelines. It is, therefore, advisable to design all animal experimentation on cellular products taking into account the following relevant nonclinical study types:

- Pharmacodynamic "proof-of-concept": Homologous animal models (i.e., animal models representative of the clinical situation and thus that provide interpretable data) should be used when possible to explore the potential clinical effect of the cellular product. Usually, a disease model is looked for first, and if not available, others are tested.
- Biodistribution: Ideally, all organs must be tested after transplantation of cells (with a safety margin of 10-fold clinical dose) into animals of two different species (one rodent, one nonrodent) and of both sexes. Of note, the testing will be dependent on the product and route of administration. These animals may also be used for environmental risk assessment and gonad tests to check for unexpected germline transmission in the case of gene therapy products. In this case, follow-up must be tied to the window of detection of transgene expression.
- Dose studies: The chosen cell dose must be based on the protocol rationale, and a dose escalation study should confirm the rationale. Toxicity studies must also be taken into account when deciding dose, and a calculation of viable/effective cellular dose in the target organ must be provided. Obviously, cell dosing is not always applicable, for example, TEPs often have a defined maximum cell load, and product application does not take into account cell dose but other parameters such as construct size, surface, etc.
- Toxicity studies: They will be performed in one species (the most relevant) and with the same route and administration method as that scheduled for the clinical trial. Unless this is not possible for practical reasons, we must find a cellular dose where toxic effects are detected and explore histopathological findings, duration, and reversibility of toxicity, as well as suitable toxicity biomarkers for our product.
- Immunogenicity and immunotoxicity studies will usually be relevant when allogeneic cells are to be used and/or if multiple dosing protocol is to be performed.
- Carcinogenicity, oncogenicity, and tumorigenicity studies will seldom be necessary, although this will depend greatly on the nature of the ATMP. For instance, tumorigenicity studies are usually required if growth factors are used in cell culture/final product and/or the product contains pluripotent or multipotent stem cells.

Many of these studies can be grouped together so that relevant evidence is obtained while ensuring best possible standards in animal welfare. In addition, the EMA has published several specific guidelines on the subject of nonclinical studies of ATMPs that should be consulted at the time of protocol design. It is also good practice, and we strongly recommend, to apply for scientific advice or protocol assistance (for orphan drugs only) from the regulatory authorities, as early in the development process as possible, and as many times as necessary throughout the course of the process.

3.2 Advanced Therapies as Medicinal Products: Manufacturing Aspects

Once we have determined that a particular product falls within the ATMP category, regardless of whether it be investigational (i.e., a drug to be used in clinical trials and not yet authorized for marketing as such), production of cells, vectors, or TEPs that will go into patients, it must comply with Good Manufacturing Practice (GMP) for medicinal products [13].

The GMP Guide is presented in two parts: basic requirements and specific annexes. GMP Part I covers all aspects in the manufacture of medicinal products, including quality assurance and risk management, and Part II deals with active substances used as starting materials [27]. In addition to Part I and II, a series of annexes providing details about specific areas of activity are included. For cell manufacturing processes, aspects of different annexes will apply (e.g., annex on sterile preparations and on biological medicinal products, among others). From a practical point of view, implementation of GMP in a cell production facility will ensure the following:

1. The existence of a quality management system
2. That there is sufficient, suitably trained personnel
3. That the premises and qualified equipment are fit for purpose and that there are separate production, quality control, and storage areas within the facilities, and cell production is being performed in a tightly controlled environment
4. That laboratory procedures are reliable and properly documented, ensuring traceability of cells and starting materials
5. That cell production operations are carried out in a controlled and reproducible (i.e., validated) manner, ensuring absence of cross-contamination
6. That the quality of cells and also the personnel, production process, and the facilities are regularly controlled
7. That the cell production facility will regularly self-inspect to monitor compliance with GMP and implement corrective measures when necessary

3.2.1   Personnel and Hygiene Needs under GMP

First and foremost, there must be sufficient qualified personnel to carry out all the tasks needed to get cells into clinical trials. All personnel should be aware of their individual responsibilities and of the GMP principles that affect them. Key posts that should be occupied by full-time personnel include the Head of Production, the Head of Quality Control, and the Qualified Person (QP) (technical director that will hold legal responsibility alongside the clinical trial Sponsor). However, this point is at the discretion of the national regulatory agencies that will often permit the doubling up of GMP responsibilities with related research, medical, or teaching duties. The heads of Production and Quality Control must be independent of each other. They should receive initial and continuing training, including hygiene instructions, as relevant to their needs. The manufacturer should provide training for all the personnel whose duties take them into production areas or into control laboratories (including the technical, maintenance, and cleaning personnel) and for other personnel whose activities could affect the quality of the product. Continuing training should also be given, and its practical effectiveness should be periodically assessed. Training records should be kept.

3.2.2   ATMP Production Facilities under GMP

Production of ATMPs will usually be performed in "cleanrooms" in which the environmental conditions (temperature and relative humidity) must be controlled, as appropriate for the intended cell culture work. Furthermore, airborne particle concentration and sterility in the working area must be tightly controlled, so that they comply with the maximum average numbers permitted for areas within each GMP "grade" (there are four such grades, termed A–D; please refer to Chapter 5, Good Manufacturing Practice compliance in the manufacture of cell-based medicines, in this book for more details).

To further comprehend just how "clean" these average particle numbers are, it must be taken into account that the generation of contamination is proportional to operator activity. A motionless person may generate about 100,000 particles $\geq 0.5\,\mu m$ per minute, and a person walking, five million particles $\geq 0.5\,\mu m$ and thousands of microbe-carrying particles per minute [28]. For this reason, only the minimum personnel required should be present in clean areas, and they should restrict their movements as much as possible.

Although the layout of GMP facilities will generally depend on the nature of the ATMP to be manufactured, Figure 5 shows a representative scheme to illustrate some considerations on GMP design that are specific for cell production facilities.

As a general principle, cell production and end-product packaging must be done in separate laminar flow hoods (GMP grade A) within a grade B environment (Figure 5(A)). Both rooms will be connected through wall-mounted, pass-through chambers with an interlock system that permits transfer of materials in and out of the cleanrooms, while avoiding contamination risks. These chambers can also be equipped with UV lamps for external sterilization of materials, if needed. Manufacturing and packaging areas will usually have positive pressure to avoid entrance of contaminants from adjacent areas. However, if virus containment is needed, the cleanroom should be negatively pressurized and adjacent to a positively pressurized "barrier" entrance room. If there is more than one door in any room, a warning or locking device is fitted to avoid simultaneous opening. Entrance to the GMP area follows a series of changing rooms where garments will be changed. Annex 1 of GMP specifies clothing required for each grade. These changing rooms also serve the purpose of gradually escalating positive pressure and air quality as the operators approach and enter the cleanroom areas, to ensure that air is not transferred from an area of higher contamination to one of lower contamination (Figure 5(A)). To further avoid contamination, it is also important that operator and material inward and outward workflows cross with each other as little as possible and only in the nonclassified areas (Figure 5(B–C)).

3.2.3 ATMP Characterization

Typical regulatory concerns with cellular components are product safety, characterization of the cells, and characterization and control of their manufacturing process. With regard to safety, cell donors must be carefully screened, and the cellular product, once expanded in the production facilities through master and working cell banks, if applicable, must be checked by several standardized tests (viability, sterility, adventitious agents, genetic stability/tumorigenicity, endotoxin, *mycoplasma* infection, etc.). Cell products will usually have to be defined as for identity, purity, potency, stability, and viability. These pharmaceutical definitions are sometimes difficult to implement in the

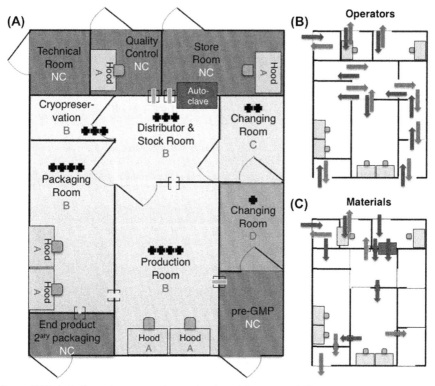

Figure 5 Technical requirements and operational workflows at an ATMP production facility under GMP. Schemes of a typical GMP facility for cell therapy and workflows of operators and materials are shown. (A) Premises will usually include nonclassified (NC, orange-colored) (dark gray in print versions) rooms as well as cleanrooms of increasing air quality (D–brown (gray in print versions), C–green (light gray in print versions), and B–blue (light gray in print versions)). Maximum air quality (A level) is usually achieved within laminar flow hoods only. Pressure (+ symbols) will increase gradually as well, to avoid contaminants entering the cleanrooms alongside the operators. GMP rules also demand that separate storage, quality control, and end-product secondary packaging areas do exist. Stock of working aliquots may be stored in clean areas. To avoid contaminants, CO_2 and liquid N_2 containers are usually left out of the cleanrooms, in a purpose-built technical room. (B) Operational workflows for personnel. Entry of personnel into the cleanrooms follows a gradient of garment changes, increasing air purity, and positive pressure. Both when entering and leaving the rooms, the operators should carry no material with them unless necessary. (C) Material flow (entry of production and exit of waste material). Materials will get into cleanrooms through autoclave, pass-through chambers, or pipes (for CO_2 and liquid N_2), and waste materials, and end product will leave them through separate pass-through chambers. Of note, operator and material workflows must cross as little as possible to avoid contamination and mistakes.

context of live cells, and they will be highly specific for the particular cell type chosen and the intended mechanism of action proposed as the rationale for the clinical trial. The manufacturing process will have to be demonstrated as aseptic (use of antibiotics is not recommended and a "media fill" validation of operational procedures must be done in advance of protocol approval) and reproducible, lot to lot consistency being of utmost importance.

As years go by, more and more protocols for manufacturing of ATMPs under GMP have been published. These should certainly be consulted since many of the requirements that agencies will ask us to comply with are already discussed in some of these publications. Suitable manufacturing solutions for our product might already be there. A small, not comprehensive sample of relevant publications for each ATMP category follows.

1. Cell therapy
   a. Facility set up [29].
   b. Isolation and expansion of hESCs [30], mesenchymal stromal cells [31–36], and cord blood cells [37].
   c. Cell encapsulation [38,39].
   d. Process scale up [40–43].
   e. Preclinical and clinical experience with mesenchymal stromal cells [44–46].
   f. Quality risk management approach [47].
   g. Information management [48].
   h. End-product shipment [49].
   i. Cell characterization assays [50].
   j. Current status of clinical trials in the field [51,52].
2. Gene therapy
   a. Facility set up [53,54].
   b. Production of plasmid DNA as a pharmaceutical [55].
   c. Production of viruses [56].
   d. Purification and characterization of adenoviral vectors [57].
   e. Purification and characterization of lentiviral vectors [58,59].
   f. Purification and characterization of retroviral vectors [60–62].
   g. Purification and characterization of AAV vectors [63–65].
   h. Nonviral vectors for gene therapy [66].
   i. Risk assessment and biosafety considerations [67,68].
   j. Current status of clinical trials in the field [69].
   k. Regulatory aspects [70].
3. Tissue engineering
   a. Bioreactor-based engineering of cartilage grafts [71].
   b. Quality and sterility analysis of cartilage transplants [72].
   c. Translating TEPs into the clinic [73].

3.3   Clinical Research with ATMPs

Before embarking on clinical trials, researchers must have approval from an Ethical Committee or Institutional Review Board (IRB) for all centers involved as well as an authorization from the national regulatory agencies of the countries where patients will be treated. To guarantee respect of human rights, to ensure data quality, and to steer clear of avoidable errors, European Directives on Good Clinical Practice (GCP) and associated guidelines must be complied with [74–86]. Likewise, and specifically for the clinical translation of stem cells, the EMA and ISSCR guidelines [77] make a good starting point. Setting up a clinical trial may be a medium- to long-term objective for many researchers in the advanced therapies field. However, it is important to keep in mind the significant amount of documentation that will be requested from the sponsor by regulatory authorities. Among other standardized forms, they will need to produce the following:

- Clinical trial protocol
- Investigator's brochure, that is, a compilation of clinical (if available) and nonclinical data on the investigational medicinal product(s) used in the clinical trial
- IMPD (termed investigational new drug—IND in the United States) that represents the main basis for approval to conduct clinical trials in Europe

The IMPD provides information on the quality, manufacture and control, nonclinical (toxicology and pharmacological tests) and clinical characteristics of the investigational medicinal product to be used in the clinical trial, including reference products and placebos. An overall risk–benefit assessment, critically analyzing the quality, nonclinical, and clinical data in relation to the potential risks and benefits of the proposed trial must also be included in the IMPD. Once the clinical trial is authorized and patient recruitment has started, the sponsor has a legal requirement to communicate to the regulatory authorities any adverse reactions. The sponsor's duties also include ensuring that there is an insurance policy in place to cover any liability, that recruitment of subjects is done after appropriate informed consent, and that approval of medicinal product batches for release conforms to specifications.

3.4 End of the Road: Marketing Authorization, Distribution, and Pharmacovigilance of ATMPs

If the regulatory bodies are satisfied that the quality, safety, and efficacy of an ATMP are sufficiently proven through successful clinical phases, a product can be granted a marketing authorization. This must be done through the centralized procedure at EMA [3], and approval would mean Europe-wide commercialization rights. For this reason, the requirements set are usually higher than those pertinent to clinical trial applications, since the number of patients to be potentially treated might be enormous for some prevalent conditions. The requisites and procedure for commercialization of ATMPs are outside the scope of this review since they will normally be relevant for pharmaceutical companies only. So far, there are five ATMPs that have successfully passed through marketing authorization at the EMA [78]:

1. An industrial TEP based on autologous chondrocytes expanded for cartilage regeneration (*ChondroCelect*)
2. Another TEP, also based on autologous chondrocytes but incorporated into a matrix (*MACI*; at present, its marketing authorization has been suspended for commercial reasons)
3. A GTMP containing the human lipoprotein lipase gene in an adeno-associated virus for the treatment of lipoprotein lipase deficiency (*Glybera*)
4. An SCTMP based on activated autologous peripheral-blood mononuclear cells for treatment of metastatic prostate cancer (*Provenge*)
5. A TEP consisting of ex vivo expanded autologous human corneal epithelial cells containing limbal stem cell for the treatment of limbal stem cell deficiency due to ocular burns (*Holoclar*). This product was the last one to receive a marketing authorization in February 2015.

Of note, all EU Member States permit exceptions to this authorization rule depending on the nature of the product, industrially prepared or otherwise. This is based on the exclusion considered by European Regulation for ATMPs [3] "which are prepared on a nonroutine basis according to specific quality standards, and used within the same Member State in a hospital under the exclusive professional responsibility of a medical practitioner, in order to comply with an individual medical prescription for a custom-made product for an individual patient." This exclusion is commonly named "hospital exemption," and it will be further explained later in this chapter.

In order to give small and medium-sized enterprises (SMEs) an incentive to conduct quality and nonclinical studies on ATMPs, a Regulation [79] came into force in 2009. Accordingly, the EMA Committee for Advanced Therapies (CAT) published a related guideline on the minimum quality and nonclinical data required for certification of ATMPs [80].

Finally, the safety of IMPs and pharmacovigilance is a key aspect of all research with ATMPs. These products are considered relatively high risk and regulatory authorities will require tight safety follow-up of ATMP-treated patients, both in clinical trials and after marketing authorization [1,16]. Once in the market, products should be consistently stored and handled as required by the marketing authorization or product specification, in accordance to Good Distribution Practice (GDP) [81], thereby maintaining the quality of the medicinal products being distributed.

4. Regional and National Institutions Supporting Cell Therapy Translational Research

Due to the innovative and complex nature and the technical specificity of cell-based medicinal products as well as the regulatory requirements for the development of translational research in this field, they are at times overwhelming for researchers and clinicians, and what is worse, they are often not successfully translated from the laboratory bench to the clinic.

Trying to fill that gap some not-for-profit organizations are promoted by regional or national governments that specifically promote the field of advanced therapies. One of the pioneering examples is the California Institute for Regenerative Medicine (CIRM [83]), which originated in 2004 when voters approved the California Stem Cell Research and Cures Initiative. CIRM was then created to fund stem cell research in the state. CIRM has quickly become a success story with an impressive list of active disease-specific projects that are reaching the clinical stage.

The Regional Government of Andalusia—having pioneered in 2003 embryonic stem cell legislation in Spain—created the Andalusian Initiative for Advanced Therapies in 2008 (IATA from the Spanish Iniciativa Andaluza en Terapias Avanzadas [84]), a publicly funded organization that gives support and training targeting researchers and clinicians needs. IATA is part of the Andalusian Public Healthcare System that offers complete health services to about 8.5 million people and comprises, among other infrastructures, 47 hospitals, around 1500 primary care centers, several research centers and institutes, a genomic and bioinformatics platform, and a Biobank storing more than 800,000 samples from patients and normal controls—including hiPS and hES cell lines. This initiative is not only focused on funding infrastructures and research projects but is also providing global support to ATMP development and translation into the clinic. IATA coordinates a network of 10 GMP facilities to manufacture gene- and cell-based therapies [85], acts as sponsor of clinical trials (24 clinical trials so far [86]), and looks for opportunities for business collaboration. IATA also organizes a master program in manufacturing of ATMPs in collaboration with the University of Granada [87].

There are many other examples of supportive organizations more focused on accelerating the commercialization of cell-based products and technologies as well as on driving the growth of the industry [88]. Some of the best known are the Center for Commercialization of Regenerative Medicine (CCRM [89]), a Canadian not-for-profit organization established in 2011, and the UK Cell Therapy Catapult [90], established in 2012.

5. European Regulation for Advanced Therapy Medicinal Products Not Intended to be Placed on the Market: Hospital Exemption and Its National Interpretation

5.1 What Does Hospital Exemption Mean?

Regulation (EC) No 1394/2007 of the European Parliament and of the Council [3] lays down specific rules concerning the authorization, supervision, and pharmacovigilance of ATMPs. Among these rules, the aforementioned Regulation establishes a centralized marketing authorization procedure for these products when they are intended to be placed on the market or industrially prepared, amending Directive 2001/83/EC [2] and Regulation (EC) No 726/2004 [91].

However, the text of the Regulation includes the following consideration regarding its scope:

"This Regulation is a lex specialis, which introduces additional provisions to those laid down in Directive 2001/83/EC. The scope of this Regulation should be to regulate advanced therapy medicinal products which are intended to be placed on the market in Member States and either prepared industrially or manufactured by a method involving an industrial process, in accordance with the general scope of the Community pharmaceutical legislation laid down in Title II of Directive 2001/83/EC. Advanced therapy medicinal products which are prepared on a non-routine basis according to specific quality standards, and used within the same Member State in a hospital under the exclusive professional responsibility of a medical practitioner, in order to comply with an individual medical prescription for a custom-made product for an individual patient, should be excluded from the scope of this Regulation whilst at the same time ensuring that relevant Community rules related to quality and safety are not undermined."

In fact, the Regulation (EC) No 1394/2007 in its article 28, point 2, incorporates an amendment to Directive 2001/83/EC of the European Parliament and of the Council of

November 6, 2001 on the community code relating to medicinal products for human use. The scope of this Directive is established in its article 2 as follows

"This Directive shall apply to medicinal products for human use intended to be placed on the market in Member States and either prepared industrially or manufactured by a method involving an industrial process."

In article 3 it indicates that the Directive *shall not apply* to certain specific medicinal products. Article 28, point 2, of the Regulation 1394/2007, introduces a further amendment to the exclusions set out in article 3, adding the following:

"Any advanced therapy medicinal product, as defined in Regulation (EC) No 1394/2007, which is prepared on a non-routine basis according to specific quality standards, and used within the same Member State in a hospital under the exclusive professional responsibility of a medical practitioner, in order to comply with an individual medical prescription for a custom-made product for an individual patient.

Manufacturing of these products shall be authorised by the competent authority of the Member State. Member States shall ensure that national traceability and pharmacovigilance requirements as well as the specific quality standards referred to in this paragraph are equivalent to those provided for at Community level in respect of advanced therapy medicinal products for which authorisation is required pursuant to Regulation (EC) No 726/2004 of the European Parliament and of the Council of March 31, 2004 laying down Community procedures for the authorisation and supervision of medicinal products for human and veterinary use and establishing a European Medicines Agency."

This article concerns what is commonly called "hospital exemption" and was included in the Regulation in recognition of the small scale and developmental nature of activity carried out in some hospitals, which calls for a degree of flexibility in the nature of regulatory requirements.

In summary, under article 28 [2] of the Regulation 1394/2007, there is an exemption from central authorization, and the Directive 2001/83/EC is not applicable to those ATMPs, which are as follows:

1. Prepared
   a. On a nonroutine basis
   b. According to specific quality standards
2. Used within the same Member State
   a. In a hospital
   b. Under the exclusive professional responsibility of a medical practitioner
3. In order to comply with an individual medical prescription
   a. For a custom-made product
   b. For an individual patient

Therefore, Directive 2001/83/EC does not apply to these products, but Member States have to ensure that the manufacture of ATMPs under hospital exemption is authorized by the competent national authority. In addition, traceability, pharmacovigilance, and specific quality standards must be equivalent to those to which ATMPs are subjected where centralized market authorization would be granted by the EMA.

The Directive does not specify what is meant by "industrial process" neither does the Regulation specify the meaning of a "custom-made product." Nevertheless, some countries have defined those terms. In the United Kingdom, the Human Tissue Authority has set definitions of the terms "custom-made" and "industrial process" [92]. Custom-made was defined as follows: "using a one off formulation or a formulation that has been tailored to the individual patient and prepared within the same hospital." "An industrial process would generally take place in an external facility and not within the same hospital." This is a very particular interpretation because the most important aspect here concerns the process (industrial or not), not the location, as the same process can take place in a facility inside or outside a hospital. For example, it may be possible to carry out the same custom-made process in a research center, a tissue bank, or even at a contract manufacturers' site. In fact, the scope of the hospital exemption considered in the Regulation 1394/2007 is irrespective of the type of manufacturer.

It is important to take into account that Regulation shall be binding in its entirety and directly applicable in all Members States, therefore its transposition into national law is not necessary. However, regulations can contain amendments of Directives that then again have to be transposed. Due to different interpretation, national transposition may result in variable or even conflicting provisions.

Article 28 [2] of the Regulation 1394/2007 is an amendment to Directive 2001/83/EC, and therefore, transposition into national law is necessary. Some European countries have already done this. The first were Finland, the United Kingdom, and Germany. Others have followed them and some others are in the process. All of them have to face some issues that arise in the interpretation of this Regulation, especially those related to the concept of nonroutine basis (e.g., small-scale production, nonroutine manufacturing procedures, and patient-specific product individually modified) and the specific quality standards (e.g., GMP and product specifications).

5.2 Nonroutine Basis

Regarding the definition of nonroutine basis, we can see different interpretations between countries since the European Commission has never specified any particular number to constitute nonroutine. Some examples are described next:

- The Medicines and Healthcare Products Regulatory Agency (MHRA), which is responsible for the regulatory arrangements under the exemption in the United Kingdom, takes the view that "it is not feasible to provide a simple numerical formula that would delineate the boundary between routine and nonroutine production" [93]. However, the agency considers that there are two main areas for consideration in determining whether preparation of a product by an operator is routine/nonroutine: whether it is the same product under consideration and the scale and frequency of the preparation of the specific product.

  "Where a number of different products are under consideration, the MHRA understands that the question of whether preparation is nonroutine should be considered separately in relation to each product prepared by that operator."

"Where a new product results from modifications to an earlier product, consideration of whether the new product is produced routinely is based on consideration of the pattern of production of that new product (and not that of the old product)."

In determining what constitutes the same product, the MHRA takes into consideration the nature of the advanced therapy medicinal product in question (product's mode of action and its intended use, as well as the manufacturing processes used to generate the final product, and any required product intermediates or product-specific starting materials, e.g., a genetically modified retrovirus used to transduce patient-specific stem cells).

"Repetition of preparation of the same product by an operator gives rise to the possibility that production of that product should be regarded as routine." The MHRA takes into account "the overall numbers of the particular product prepared by the operator, the regularity/frequency of production, and the time period over which the preparation of that product has become established."

- The case of the Netherlands is an example of a country that has chosen a very concrete way to define nonroutine basis. The competent authority is the Dutch Health Care Inspectorate. Under the hospital exemption, the infusion of one product for a maximum of five patients and fewer than 10 patients a year [94] is allowed.
- Germany, Finland, France, Spain, Portugal, and Italy have not yet established any number to define nonroutine. They consider the concept of hospital exemption more flexibly. Concretely, Germany, which has implemented the hospital exemption into the German Medicinal Products Act [95], has a legal definition for ATMPs prepared on a nonroutine basis as those "medicines:
  - which are manufactured in small quantities, and in the case of which, based on a routine manufacturing procedure, variations in the procedure which are medically justified for an individual patient are carried out, or
  - which have not yet been manufactured in sufficient quantities so that the necessary data to enable a comprehensive assessment are not yet available."

  The Website of the Paul-Ehrlich-Institut (PEI), the higher federal authority, set up a decision tree (Figure 6) available to inform the decision as to when the hospital exemption rule applies under the German Medicinal Products Act [96].

5.3 Quality Standards and Other Requirements

Most countries that have regulated this hospital exemption require GMP as the quality standard applicable under the hospital exemption scheme. Nevertheless, there are some differences between countries. For example, in the case of the United Kingdom, a QP is not required [97]. In Germany, "person identity" of the manufacturer is not necessary. In the Netherlands [94], France [98], and Spain [99], although GMP is required, there is some kind of flexibility.

The regulation establishes the same requirements in terms of pharmacovigilance and traceability for ATMPs independent of whether they are granted centralized market authorization by the EMA or in the case of application for hospital exemption.

Below, we specify some additional information about the requirements for applying for the hospital exemption in some European countries.

- Germany, through the PEI, gives specific authorization—Section 4b of the German Medicinal Products Act [95]—for specific products or indications. The authorization is granted

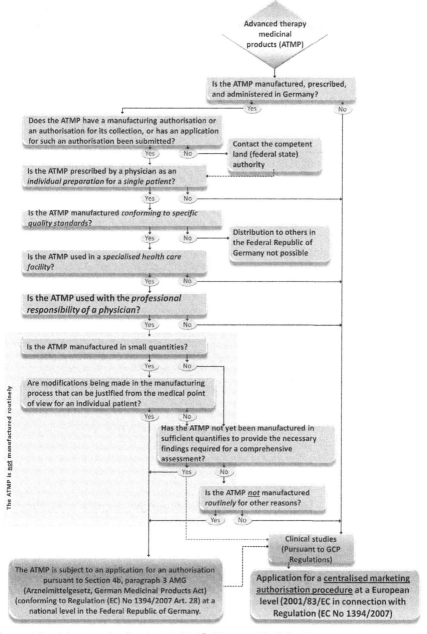

Figure 6 Decision tree for Section 4b AMG (German Medicinal Products Act). Paul-Ehrlich-Institut. Available from: http://www.pei.de/SharedDocs/Downloads/EN/pu/innova-tion-office/decision-tree-4b-amg.pdf?__blob=publicationFile&v=1 [accessed on 06.03.15].

not necessarily to a hospital. However, the authorized use is restricted to a specialized health care facility with proven specialization. The authorization holder has the obligation to report to the PEI the amount of preparation and knowledge required to enable the comprehensive assessment to take place. Finally, the authorization may be withdrawn or revoked by the PEI. As for every investigational medicinal product, Manufacturing Authorization is required but is issued by the competent authority of respective Laender. Therefore, in the case of hospital exemption in Germany, a manufacturing authorization is necessary as well as provisions for traceability and pharmacovigilance equivalent to those for other ATMP.

- In the case of Finland, the hospital exemption has been transposed into the Finnish legislation in the Medicines Act, Section 15c [100] (general requirements) and in the Administrative Regulation 3/2009 [101] (technical requirements). The nonindustrial manufacture of ATMPs "is subject to licence granted by the Finnish Medicines Agency (Fimea). The licence may be granted for the manufacture of a medicinal product by prescription from a physician for the individual treatment of a particular patient in a hospital. The licence may incorporate conditions pertaining to the preparation, release, traceability and use of the medicinal product or required for medicinal product safety." The application for hospital exemption manufacturing must include the following: identification of the manufacturer, description of the ATMP and product-specific quality requirements, information concerning prescribing and the doctor responsible for the patient care, description of the manufacturing process, persons responsible for the manufacturing process, manufacturing personnel; competence of the manufacturing personnel, general description concerning the quality system, manufacturing premises, critical equipment and material for the quality of the ATMP, procedure to confirm traceability requirements, procedure for serious adverse events, procedure for pharmacovigilance, declaration concerning registered personal data, and an ethical assessment and environmental effects assessment (specifically for GTMPs). Manufacturing should comply with GMP principles, and it is necessary to send an annual report to Fimea [101].
- In the United Kingdom, the MHRA has published guidance [93] that sets out the requirements relating to GMP, pharmacovigilance, traceability, and patient information under the hospital exemption. "In the UK, a manufacturer will be required to obtain a manufacturer's licence from the MHRA. The licence will authorize the manufacture of particular categories of ATMPs (gene therapy, somatic cell therapy, or tissue engineered product) rather than individual products in line with current manufacturer's licensing arrangements. ATMPs made and used under the exemption must comply with the principles of GMP." The MHRA has also included guidance on other requirements not specified in the Regulation regarding to labeling, package leaflet, advertising, and ethical issues. Manufacturers operating under the hospital exemption are required to make an annual return to the MHRA.
- In Spain [99] and Portugal [102], the licence is given to hospitals. In Spain, this licence is irrespective of the manufacturer; therefore, if several Spanish hospitals were interested in using an ATMP produced by a single manufacturer, each hospital should submit a dossier equivalent to a Common Technical Document.
- Italy [103] is about to publish a Ministerial Decree regulating the hospital exemption. The Italian Medicines Agency (AIFA, Agenzia Italiana del Farmaco) is the competent authority to authorize the manufacture of ATMPs under the hospital exemption as well as their use. The hospital exemption is only granted for public institutions, requiring authorization of the manufacturing facility, according GMP rules, and authorization of the hospital (only for public hospitals, university hospitals, or biomedical research institutes).

5.4   When to Apply for the Hospital Exemption

Perhaps the most important issue regarding hospital exemption is the situation in which the hospital exemption applies. The conditions that must be complied with are clear in the Regulation: ATMPs "prepared on a nonroutine basis according to specific quality standards, and used within the same Member State in a hospital under the exclusive professional responsibility of a medical practitioner, in order to comply with an individual medical prescription for a custom-made product for an individual patient" but not the circumstances in which applications should be made.

Is it possible, under this hospital exemption, to understand that it could apply before starting a clinical trial as a "proof-of-concept" in human beings? Or instead of a clinical trial? If so, should a positive decision of an ethics committee and patient insurance be in place under the hospital exemption? Is it an alternative way to compassionate use? Or, after finishing clinical trials to introduce the therapy in question as standard of care instead of a marketing approval? Or in cases where the development of products began before they were considered as medicinal products?

Some countries have attempted to answer this question, while others have not. But the answer to this question is not easy, as the Regulation does not specify that point, making different interpretations possible.

- For example, the guidance on the United Kingdom's arrangements under the hospital exemption scheme does not apply to ATMPs that will be authorized under the ATMP Regulation, for which the centralized marketing authorization procedure will apply, nor does it apply to ATMPs supplied as investigational ATMPs for use in a clinical trial.

In addition to that, the guidance incorporates a distinction between hospital exemption and specials (Table 1). "Although the two schemes are legally distinct, there are some apparent

Table 1 Summary of Some of the Main Differences in Scope between the Hospital Exemption and "Specials" Schemes in the United Kingdom

Hospital Exemption	The "Specials" Scheme
The ATMP must be prepared and used in the same EU Member State.	Products meeting the requirements of the scheme can be manufactured in the UK or imported to the UK.
The ATMP must be commissioned by a medical practitioner.	Products can be prescribed by doctors, dentists, and supplementary prescribers.
The ATMP must be custom made to meet an individual prescription and preparation must be on a "non-routine basis."	There is a special needs test (interpreted to mean the absence of a pharmaceutically equivalent and available licenced product).
The ATMP must be used in a hospital.	There is no stipulation as to location.

Available from: https://www.gov.uk/government/uploads/system/uploads/attachment_data/file/397738/Guidance_on_the_UK_s_arrangements_under_the_hospital_exemption_scheme.pdf [accessed on 06.03.15].

similarities between the kind of activities falling within the hospital exemption and the UK "specials" scheme. Products made or supplied under either scheme are referred to as "unlicenced" since there is no product licence (marketing authorization). However, each site will need to hold a manufacturer's licence of a type specific to the scheme. It should be noted that a QP is not required for either scheme. The UK "specials" scheme, including the linked import notification scheme, permits doctors and certain other prescribers to commission an unlicenced relevant medicinal product to meet the special needs of individual patients. In principle, this latter scheme would be available for ATMPs as for any other category of medicinal product. The MHRA expects that there may in practice be a variety of situations in which small-scale production of an unlicenced ATMP is envisaged to meet requests made by a prescriber. In these circumstances operators will need to consider carefully which of the two schemes, (if either), is applicable."

In summary, the guidance specifies clearly when the hospital exemption does not apply, but it is not so clear regarding when it does. It may be possible that its purpose is to foster early stage product development.

- In Finland, the hospital exemption is understood, in some way, as a prior phase to clinical trials. Fimea has tried to tailor the quality requirements to the same level as in the first-in-man clinical trials, because then the applicants have quite easy access to the clinical trial pathway later on. The requirements are at the same level as clinical trials for quality but not for nonclinical data (not required under hospital exemption). The idea is to allow small-scale clinical use while nonclinical studies are carried out to facilitate a later clinical trial.
- In the Netherlands, the hospital exemption is applied for patients ineligible for a clinical trial (as in compassionate use) or in cases in which the product required falls outside of the specifications.
- In Italy, the hospital exemption can be granted when there is not any other alternative therapy or in case of emergency or life-threatening conditions.
- The requirements regarding quality and nonclinical information in Germany depend on the nature of the individual product, already available clinical data, and its medical need. However, at least data comparable to those of early investigational products are expected to be available. Thus, in this scenario, it might be conceivable that the hospital exemption might be applicable to facilitate or accompany a clinical trial, but also for conveying products in a preliminary "nonroutine status" on their way to centralized marketing approval.
- However, in Spain, it is only possible to apply for hospital exemption when efficacy and safety have been demonstrated, and quality, nonclinical, and clinical data must be provided. Here hospital exemption is considered as an alternative to marketing authorization for products nonindustrially manufactured and not intended to be marketed, but not as an alternative to clinical trials.

On the subject of the duration of the authorization under the hospital exemption, there is great variability between countries. At one extreme is the Netherlands, where the product-specific licence lasts for 10 batches or for one year and, at the other, Germany, where the licence is also product specific but where there is no specific period. In the case of Finland, the licence for nonindustrial manufacture of the ATMPs may be granted for a fixed or indefinite term. In Portugal, it is granted for one year, being renewable. And in Spain, authorization is initially given for three years, and for five years in successive renewals.

6. Conclusions

ATMP development is a long and risky process due to the fast pace of advancement of the science in these fields, which are currently booming. Moreover, the regulation is continuously being adapted, and thus, some degree of uncertainty will always be present while products are in the pathway to market authorization. A major conclusion of this work is that large, multidisciplinary teams with the required expertise must be assembled to be able to translate an ATMP bench to bedside.

As we have seen, hospital exemption introduced into the Regulation 1394/2007 made the Directive 2001/83 inapplicable for ATMPs "prepared on a nonroutine basis according to specific quality standards, and used within the same Member State in a hospital under the exclusive professional responsibility of a medical practitioner, in order to comply with an individual medical prescription for a custom-made product for an individual patient."

At the same time, the Regulation established that Member States have to ensure that the manufacture of ATMPs under the hospital exemption is authorized by the competent authority and that traceability, pharmacovigilance, and specific quality standards must be equivalent to those for ATMPs industrially manufactured.

After reviewing how this amendment of the Directive 2001/83 has been transposed into different national legislations, we can conclude that there are some important differences between countries, not only concerning the requirements to apply for the hospital exemption but also when to apply for it. Some stakeholders ask for more harmonization in the interpretation of hospital exemption within the European Union. However, hospital exemption was provided just to allow some flexibility and to enable different Member States to fit their individual circumstances to ATMP Regulation. SMEs and not-for-profit organizations (mainly universities and public hospitals) are leading the clinical development of ATMPs in Europe, with great variability among countries. In Spain and Italy, the role of industry has been minimal; while in the United Kingdom, Germany, France, Sweden, and Denmark, it has been quite important. These differences have probably been instrumental in the way national authorities have implemented hospital exemption [104].

References

[1] Directive 2001/83/EC of the European Parliament and of the Council of 6 November 2001 on the community code relating to medicinal products for human use. Off J Eur Communities 28.11.2001;L 311:67–128.
[2] Commission directive 2009/120/EC of 14 September 2009 amending directive 2001/83/EC of the European Parliament and of the Council on the community code relating to medicinal products for human use as regards advanced therapy medicinal products. Off J Eur Union 15.9.2009;L 243:3–12.
[3] Regulation (EC) no 1394/2007 of the European Parliament and of the Council of 13 November 2007 on advanced therapy medicinal products and amending Directive 2001/83/EC and Regulation (EC) No 726/2004. Off J Eur Union 10.12.2007;L 324:121–37.

[4] Decision tree for classification of medicinal products as ATMP. Paul-Ehrlich-Institut. Available at: http://www.pei.de/cln_101/nn_2107932/SharedDocs/Downloads/EN/pu/innovation-office/decision-tree-atmp,templateId=raw,property=publicationFile.pdf/decision-tree-atmp.pdf [accessed 06.03.15].

[5] European Medicines Agency. Committee for Advanced Therapies (CAT). Reflection paper on classification of advanced therapy medicinal products. 2015. EMA/CAT/600280/2010 Rev.1. EMA web site [online] http://www.ema.europa.eu/docs/en_GB/document_library/Scientific_guideline/2015/06/WC500187744.pdf.

[6] Cuende N, Rico L, Herrera C. Bone marrow mononuclear cells for the treatment of ischemic syndromes: medicinal product or cell transplantation? Stem Cells Transl Med 2012;1:403–8.

[7] Cuende N, Herrera C, Keating A. When the best is the enemy of the good: the case of bone-marrow mononuclear cells to treat ischemic syndromes. Haematologica March 2013;98(3):323–4.

[8] Directive 2002/98/EC of the European Parliament and of the Council of 27 January 2003 setting standards of quality and safety for the collection, testing, processing, storage and distribution of human blood and blood components and amending Directive 2001/83/EC. Off J Eur Union 8.3.2003;L 33:30–40.

[9] Commission Directive 2004/33/EC of 22 March 2004 implementing Directive 2002/98/EC of the European Parliament and of the Council as regards certain technical requirements for blood and blood components. Off J Eur Union 30.3.2004;L 91:25–39.

[10] Commission Directive 2005/62/EC of 30 September 2005 implementing directive 2002/98/EC of the European Parliament and of the Council as regards community standards and specifications relating to a quality system for blood establishments. Off J Eur Union 1.10.2005;L 256:41–8.

[11] Directive 2004/23/EC of the European Parliament and of the Council of 31 March 2004 on setting standards of quality and safety for the donation, procurement, testing, processing, preservation, storage and distribution of human tissues and cells. Off J Eur Union 7.4.2004;L 102:48–58.

[12] Commission Directive 2006/17/EC of 8 February 2006 implementing Directive 2004/23/EC of the European Parliament and of the Council as regards certain technical requirements for the donation, procurement and testing of human tissues and cells. Off J Eur Union 9.2.2006;L 38:40–52.

[13] Commission Directive 2003/94/EC of 8 October 2003 laying down the principles and guidelines of good manufacturing practice in respect of medicinal products for human use and investigational medicinal products for human use. J Eur Union 14.10.2003; L 262:22–6.

[14] Commission directive 2006/86/EC of 24 October 2006 implementing Directive 2004/23/EC of the European Parliament and of the Council as regards traceability requirements, notification of serious adverse reactions and events and certain technical requirements for the coding, processing, preservation, storage and distribution of human tissues and cells. Off J Eur Union 25.10.2006;L 294:32–50.

[15] Cuende N, Izeta A. Clinical translation of stem cell therapies: a bridgeable gap. Cell Stem Cell 04.06.2010;6:508–12.

[16] European Medicines Agency. Guideline on the risk-based approach according to annex I, part IV of Directive 2001/83/EC applied to advanced therapy medicinal products. EMA/CAT/CPWP/686637/2011. 11.02.2013.

[17] European Directorate for the Quality of Medicines and Healthcare (EDQM). European pharmacopoeia. 6th ed. 2010 (6.8).

[18] European Medicines Agency. Guideline on human cell-based medicinal products. EMEA/CHMP/410869/2006. 21.05.2008.

[19] European Medicines Agency. Reflection paper on stem cell-based medicinal products. EMA/CAT/571134/2009. 14.01.2011.

[20] European Medicines Agency. Reflection paper on clinical aspects related to tissue engineered products. EMA/CAT/573420/2009. 19.09.2014.

[21] http://www.ema.europa.eu/ema/index.jsp?curl=pages/regulation/general/general_content_000405.jsp&mid=WC0b01ac058002958a.

[22] http://www.ema.europa.eu/ema/index.jsp?curl=pages/regulation/general/general_content_000410.jsp&murl=menus/regulations/regulations.jsp&mid=WC0b01ac-058002958d&jsenabled=true.

[23] European Medicines Agency. Guideline on quality, non-clinical and clinical aspects of medicinal products containing genetically modified cells. EMA/CHMP/GTWP/671639/2008. 13.04.2012.

[24] Commission Directive 2003/63/EC of 25 June 2003 amending directive 2001/83/EC of the European Parliament and of the Council on the community code relating to medicinal products for human use. Off J Eur Union 27.6.2003;L 159:46–94.

[25] Directive 2004/9/EC of the European Parliament and of the Council of 11 February 2004 on the inspection and verification of good laboratory practice (GLP). Off J Eur Union 20.2.2004;L 50:28–43.

[26] Directive 2004/10/EC of the European Parliament and of the Council of 11 February 2004 on the harmonisation of laws, regulations and administrative provisions relating to the application of the principles of Good Laboratory Practice and the verification of their applications for tests on chemical substances. Off J Eur Union 20.2.2004;L 5:44–59.

[27] European Commission. EudraLex. guidelines to good manufacturing practice medicinal products for human and veterinary use, vol. 4. 2012.

[28] Whyte W. Cleanroom disciplines. In: Cleanroom technology – fundamentals of design, testing and operation. Chichester, UK: John Wiley and Sons; 2001.

[29] Arjmand B, Emami-Razavi SH, Larijani B, Norouzi-Javidan A, Aghayan HR. The implementation of tissue banking experiences for setting up a cGMP cell manufacturing facility. Cell Tissue Bank 2012;13:587–96.

[30] Unger C, Skottman H, Blomberg P, Dilber MS, Hovatta O. Good Manufacturing Practice and clinical-grade human embryonic stem cell lines. Hum Mol Genet 2008;17:R48–53.

[31] Astori G, Soncin S, Lo Cicero V, Siclari F, Surder D, Turchetto L, et al. Bone marrow derived stem cells in regenerative medicine as advanced therapy medicinal products. Am J Transl Res 2010;2:285–95.

[32] Bieback K, Kinzebach S, Karagianni M. Translating research into clinical scale manufacturing of mesenchymal stromal cells. Stem Cells Int 2010. http://dx.doi.org/10.4061/2010/193519.

[33] Bourin P, Sensebe L, Planat-Benard V, Roncalli J, Bura-Riviere A, Casteilla L. Culture and use of mesenchymal stromal cells in phase I and II clinical trials. Stem Cells Int 2010. http://dx.doi.org/10.4061/2010/503593.

[34] Grisendi G, Anneren C, Cafarelli L, Sternieri R, Veronesi E, Cervo GL, et al. GMP-manufactured density gradient media for optimized mesenchymal stromal/stem cell isolation and expansion. Cytotherapy 2010;12:466–77.

[35] Sensebe L, Bourin P, Tarte K. Good Manufacturing Practices production of mesenchymal stem/stromal cells. Hum Gene Ther 2011;22:19–26.

[36] Torre ML, Lucarelli E, Guidi S, Ferrari M, Alessandri G, De Girolamo L, et al. Ex vivo expanded mesenchymal stromal cell minimal quality requirements for clinical application. Stem Cells Dev 2015;24:677–85.

[37] Aktas M, Buchheiser A, Houben A, Reimann V, Radke T, Jeltsch K, et al. Good Manufacturing Practice-grade production of unrestricted somatic stem cell from fresh cord blood. Cytotherapy 2010;12:338–48.

[38] van Zanten J, de Vos P. Regulatory considerations in application of encapsulated cell therapies. Adv Exp Med Biol 2010;670:31–7.

[39] Villani S, Marazzi M, Bucco M, Faustini M, Klinger M, Gaetani P, et al. Statistical approach in alginate membrane formulation for cell encapsulation in a GMP-based cell factory. Acta Biomater 2008;4:943–9.

[40] Ausubel LJ, Lopez PM, Couture LA. GMP scale-up and banking of pluripotent stem cells for cellular therapy applications. Methods Mol Biol 2011;767:147–59.

[41] Justice C, Brix A, Freimark D, Kraume M, Pfromm P, Eichenmueller B, et al. Process control in cell culture technology using dielectric spectroscopy. Biotechnol Adv 2011;29:391–401.

[42] Ratcliffe E, Thomas RJ, Williams DJ. Current understanding and challenges in bioprocessing of stem cell-based therapies for regenerative medicine. Br Med Bull 2011;100:137–55.

[43] Sharma S, Raju R, Sui S, Hu WS. Stem cell culture engineering – process scale up and beyond. Biotechnol J 2011;6:1317–29.

[44] Sensebe L, Krampera M, Schrezenmeier H, Bourin P, Giordano R. Mesenchymal stem cells for clinical application. Vox Sang 2009;98:93–107.

[45] Tolar J, Le Blanc K, Keating A, Blazar BR. Concise review: hitting the right spot with mesenchymal stromal cells. Stem Cells 2010;28:1446–55.

[46] Veronesi E, Murgia A, Caselli A, Grisendi G, Piccinno MS, Rasini V, et al. Transportation conditions for prompt use of ex vivo expanded and freshly harvested clinical-grade bone marrow mesenchymal stromal/stem cells for bone regeneration. Tissue Eng Part C 2014;20:239–51.

[47] Lopez F, Di Bartolo C, Piazza T, Passannanti A, Gerlach JC, Gridelli B, et al. A quality risk management model approach for cell therapy manufacturing. Risk Anal 2010;30:1857–71.

[48] Russom D, Ahmed A, Gonzalez N, Alvarnas J, Digiusto D. Implementation of a configurable laboratory information management system for use in cellular process development and manufacturing. Cytotherapy 2012;14:114–21.

[49] Whiteside TL, Griffin DL, Stanson J, Gooding W, McKenna D, Sumstad D, et al. Shipping of therapeutic somatic cell products. Cytotherapy 2011;13:201–13.

[50] Carmen J, Burger SR, McCaman M, Rowley JA. Developing assays to address identity, potency, purity and safety: cell characterization in cell therapy process development. Regen Med 2012;7:85–100.

[51] Swan M. Steady advance of stem cell therapies. Report from the 2011 world stem cell summit, Pasadena, California, October 3–5. Rejuvenation Res 2011;14:699–704.

[52] Trounson A, Thakar RG, Lomax G, Gibbons D. Clinical trials for stem cell therapies. BMC Med 2011;9:52.

[53] Alici E, Blomberg P. GMP facilities for manufacturing of advanced therapy medicinal products for clinical trials: an overview for clinical researchers. Curr Gene Ther 2010;10:508–15.

[54] Cohen-Haguenauer O, Creff N, Cruz P, Tunc C, Aiuti A, Baum C, et al. Relevance of an academic GMP Pan-European vector infra-structure (PEVI). Curr Gene Ther 2011;10:414–22.

[55] Schleef M, Blaesen M. Production of plasmid DNA as a pharmaceutical. Methods Mol Biol 2009;542:471–95.

[56] Vicente T, Peixoto C, Carrondo MJ, Alves PM. Virus production for clinical gene therapy. Methods Mol Biol 2009;542:447–70.

[57] Eglon MN, Duffy AM, O'Brien T, Strappe PM. Purification of adenoviral vectors by combined anion exchange and gel filtration chromatography. J Gene Med 2009; 11:978–89.

[58] Merten OW, Charrier S, Laroudie N, Fauchille S, Dugue C, Jenny C, et al. Large-scale manufacture and characterization of a lentiviral vector produced for clinical ex vivo gene therapy application. Hum Gene Ther 2011;22:343–56.

[59] Witting SR, Li LH, Jasti A, Allen C, Cornetta K, Brady J, et al. Efficient large volume lentiviral vector production using flow electroporation. Hum Gene Ther 2011;23:243–9.

[60] Carrondo M, Panet A, Wirth D, Coroadinha AS, Cruz P, Falk H, et al. Integrated strategy for the production of therapeutic retroviral vectors. Hum Gene Ther 2011;22:370–9.

[61] Herbst F, Ball CR, Zavidij O, Fessler S, Schmidt M, Veelken H, et al. 10-year stability of clinical-grade serum-free gamma-retroviral vector-containing medium. Gene Ther 2011;18:210–2.

[62] van der Loo JC, Swaney WP, Grassman E, Terwilliger A, Higashimoto T, Schambach A, et al. Scale-up and manufacturing of clinical-grade self-inactivating gamma-retroviral vectors by transient transfection. Gene Ther 2011;19:246–54.

[63] Allay JA, Sleep S, Long S, Tillman DM, Clark R, Carney G, et al. Good Manufacturing Practice production of self-complementary serotype 8 adeno-associated viral vector for a hemophilia B clinical trial. Hum Gene Ther 2011;22:595–604.

[64] Mitchell AM, Nicolson SC, Warischalk JK, Samulski RJ. AAV's anatomy: roadmap for optimizing vectors for translational success. Curr Gene Ther 2010;10:319–40.

[65] Wright JF, Wellman J, High KA. Manufacturing and regulatory strategies for clinical AAV2-hRPE65. Curr Gene Ther 2010;10:341–9.

[66] Yin H, Kanasty RL, Eltoukhy AA, Vegas AJ, Dorkin JR, Anderson DG. Non-viral vectors for gene-based therapy. Nat Rev Genet 2014;15:541–55.

[67] Bamford KB, Wood S, Shaw RJ. Standards for gene therapy clinical trials based on pro-active risk assessment in a London NHS Teaching Hospital Trust. QJM 2005;98:75–86.

[68] Pauwels K, Gijsbers R, Toelen J, Schambach A, Willard-Gallo K, Verheust C, et al. State-of-the-art lentiviral vectors for research use: risk assessment and biosafety recommendations. Curr Gene Ther 2009;9:459–74.

[69] Bamford KB. Clinical trials of GMP products in the gene therapy field. Methods Mol Biol 2011;737:425–42.

[70] Laurencot CM, Ruppel S. Regulatory aspects for translating gene therapy research into the clinic. Methods Mol Biol 2009;542:397–421.

[71] Santoro R, Olivares AL, Brans G, Wirz D, Longinotti C, Lacroix D, et al. Bioreactor based engineering of large-scale human cartilage grafts for joint resurfacing. Biomaterials 2010;31:8946–52.

[72] Pudlas M, Koch S, Bolwien C, Walles H. Raman spectroscopy as a tool for quality and sterility analysis for tissue engineering applications like cartilage transplants. Int J Artif Organs 2010;33:228–37.

[73] Bayon Y, Vertès AA, Ronfard V, Egloff M, Snykers S, Salinas GF, et al. Translating cell-based regenerative medicines from research to successful products: challenges and solutions. Tissue Eng Part B Rev 2014;20:246–56.

[74] European Medicines Agency. ICH topic e 6 (R1) guideline for Good Clinical Practice. Note for guidance on Good Clinical Practice. CPMP/ICH/135/95. July 2002.

[75] Directive 2001/20/EC of the European Parliament and of the Council of 4 April 2001 on the approximation of the laws, regulations and administrative provisions of the Member States relating to the implementation of Good Clinical Practice in the conduct of clinical trials on medicinal products for human use. Off J Eur Communities 1.5.2001;L 121:34–44.

[76] Commission Directive 2005/28/EC of 8 April 2005 laying down principles and detailed guidelines for Good Clinical Practice as regards investigational medicinal products for human use, as well as the requirements for authorization of the manufacturing or importation of such products. Off J Eur Union 9.4.2005;L 91:13–9.

[77] International Society for Stem Cell Research (ISSCR). Guidelines for the clinical translation of stem cells. Curr Protoc Stem Cell Biol 2009.

[78] European Medicines Agency. Committee for advanced therapies (CAT): CAT monthly report of application procedures, guidelines and related documents on advanced therapies. January 2015 meeting. 2015. EMA/CAT/45673/2015. EMA web site [online] http://www.ema.europa.eu/docs/en_GB/document_library/Committee_meeting_report/2015/01/WC500181069.pdf.

[79] Commission Regulation (EC) No 668/2009 of 24 July 2009 implementing regulation (EC) no 1394/2007 of the European Parliament and of the Council with regard to the evaluation and certification of quality and non-clinical data relating to advanced therapy medicinal products developed by micro, small and medium-sized enterprises. Off J Eur Union 25.7.2009;L 194:7–10.

[80] European Medicines Agency. Scientific guideline on the minimum quality and non-clinical data for certification of advanced therapy medicinal products. EMEA/CAT/486831/2008/corr. 15.10.2010.

[81] Council Directive 92/25/EEC of 31 March 1992 on the wholesale distribution of medicinal products for human use. Off J 30.04.1992;L 113:1–4.

[82] http://www.juntadeandalucia.es/terapiasavanzadas/index.php/en/2014-04-14-08-40-24/regulation.

[83] https://www.cirm.ca.gov/.

[84] http://www.juntadeandalucia.es/terapiasavanzadas/index.php/en/.

[85] http://www.juntadeandalucia.es/terapiasavanzadas/index.php/en/2014-04-14-08-40-24/gmp-laboratories.

[86] http://www.juntadeandalucia.es/terapiasavanzadas/images/clinical_trials.pdf.

[87] http://www.atmp-masterinmanufacturing.com/.

[88] Schachter B. Therapies of the state. Nat Biotechnol 2014;32:736–41.

[89] http://www.ccrm.ca/.

[90] https://ct.catapult.org.uk/.

[91] Regulation (EC) No 726/2004 of the European Parliament and of the Council of 31 March 2004 laying down community procedures for the authorization and supervision of medicinal products for human and veterinary use and establishing a European Medicines Agency. Off J Eur Union 30.04.2004;L 136:1–33.

[92] ATMP definitions. Human tissue authority. Available from: https://www.hta.gov.uk/sites/default/files/ATMP_definitions_200901191224.pdf [accessed 06.03.15].

[93] Guidance on "non routine". Medicines and healthcare products regulatory agency (MHRA). Available from: https://www.gov.uk/government/uploads/system/uploads/attachment_data/file/397739/Non-routine_guidance_on_ATMPs.pdf [accessed 06.03.15].

[94] Procedure voor het verkrijgen van een hospital exemption voor ATMPs (versie April 2011). Inspectie voor de Gezondheidszorg. Available from: http://www.igz.nl/Images/Procedure%20ATMP%20huisstijl%20rev%20april%202011_tcm294-283446.pdf [accessed 06.03.15].

[95] Medicinal Products Act (the drug law) (Arzneimittelgesetz – AMG) of the Federal Republic of Germany (non-official translation). Available from: http://www.gmp-compliance.org/guidemgr/files/AMG%20ENGLISCH%202006.PDF [accessed 06.03.15].

[96] Decision tree for Section 4b AMG (German medicinal products act). Paul-Ehrlich-Institut. Available from: http://www.pei.de/SharedDocs/Downloads/EN/pu/innovation-office/decision-tree-4b-amg.pdf?__blob=publicationFile&v=1 [accessed 06.03.15].

[97] Guidance on the UK's arrangements under the hospital exemption scheme. Medicines and Healthcare Products Regulatory Agency (MHRA). Available from: https://www.gov.uk/government/uploads/system/uploads/attachment_data/file/397738/Guidance_on_the_UK_s_arrangements_under_the_hospital_exemption_scheme.pdf [accessed 06.03.15].

[98] LOI n° 2011-302 du 22 mars 2011 portant diverses dispositions d'adaptation de la législation au droit de l'Union européenne en matière de santé, de travail et de communications électroniques (not available in English). Available from: http://www.legifrance.gouv.fr/jopdf/common/jo_pdf.jsp?numJO=0&dateJO=20110323&numTexte=6&pageDebut=05186&pageFin=05193 [accessed 06.03.15].

[99] Real Decreto 477/2014, de 13 de junio, por el que se regula la autorización de medicamentos de terapia avanzada de fabricación no industrial (not available in English). Available from: https://www.boe.es/boe/dias/2014/06/14/pdfs/BOE-A-2014-6277.pdf [accessed 06.03.15].

[100] Medicines act 395/1987 unofficial translation; amendments up to 1340/2010 included. FIMEA. Available from: http://www.fimea.fi/instancedata/prime_product_julkaisu/fimea/embeds/fimeawwwstructure/18580_Laakelaki_englanniksi_paivitetty_5_2011.pdf [accessed 06.03.15].

[101] Administrative regulation 3/2009. Pitkälle kehitetyssä terapiassa käytettävien lääkkeiden valmistaminen yksittäisen potilaan käyttöön FIMEA. (not available in English) Available from: http://www.fimea.fi/instancedata/prime_product_julkaisu/fimea/embeds/fimeawwwstructure/17048_Pitkalle_kehitetyssa_terapiassa_maarays_3_2009.pdf [accessed 06.03.15].

[102] Portaria n.° 138/2014 de 7 de julho. Diário da República, 1.ª série, N.° 128, 7 de julho de 2014 (not available in English). Available from: http://dre.pt/pdf1s/2014/07/12800/0369603702.pdf [accessed 06.03.15].

[103] Agenzia Italiana del Farmaco. Firmato Decreto Ministeriale che disciplina la preparazione e l'utilizzo dei medicinali per terapie avanzate "su base non ripetitiva". Available from: http://www.agenziafarmaco.gov.it/it/content/firmato-decreto-ministeriale-che-disciplina-la-preparazione-e-l'utilizzo-dei-medicinali-te-0 [accessed 06.03.15].

[104] Cuende N, Boniface C, Bravery C, Forte M, Giordano R, Hildebrandt M, et al. Legal and regulatory affairs committee—Europe, international society for cellular therapy. The puzzling situation of hospital exemption for advanced therapy medicinal products in Europe and stakeholders' concerns. Cytotherapy December 2014;16(12):1597–600.

Nonclinical Studies for Cell-Based Medicines

Michaela Sharpe, Giulia Leoni, Jacqueline Barry, Rindi Schutte, Natalie Mount
Cell Therapy Catapult Ltd, Guys Hospital, London, UK

Chapter Outline

1.   Introduction

Cellular therapy is based on the concept of modulation or repair of function through the administration of cells. Cells have therapeutic potential that is distinct from small molecules and biologics in that they have the ability to respond and adapt to environmental signals, to home to specific diseased tissues, and to execute complex responses in a regulated manner. Although there has recently been an acceleration in the field, the concept of cellular therapies is not new. The first successful hematopoietic stem cell (HSC) transplantation was recorded in 1968 and is now an accepted treatment for patients whose immune system is defective or has been damaged [1]. The first licensed products [2] and the recent pioneering trials, particularly in the field of cellular immunotherapies [3,4], have highlighted the potential for such therapies to revolutionize the field of medicine.

There are two principle categories of cell therapy products from a regulatory perspective in the European Union (EU): minimally manipulated products that perform the same function in the recipient as in the donor [5] and products designated as advanced therapy medicinal products (ATMPs) that have undergone manipulation and/or are designed to perform a different function in the recipient than that performed in the donor [6]. With a minimally manipulated cell therapy product, the cells are isolated from the donor and minimally processed before returning to the recipient. For structural cells (e.g., cells obtained from skin, adipose tissue, blood vessel, bone, cartilage, amniotic membrane), the processing should not alter the original relevant characteristics of the tissue relating to the tissue's utility for reconstruction, repair, or replacement, and for nonstructural tissues, the processing should not alter the relevant biological characteristics of cells.

The ATMP regulation [7] extended the definition of medicinal products (as defined in Ref. [8]), including somatic cell therapy and gene therapy medicinal products to include a definition of a tissue engineered product. A cell- or tissue-derived medicinal product can be considered engineered if it contains or consists of cells or tissues that have either been subject to "substantial manipulation" or that are "not intended to be used for the same essential function(s)" in the recipient as in the donor. The cell therapy is also presented as "having properties for treating or preventing disease" in patients. The ATMP definition, hence, includes the following groups of product: a gene therapy medicinal product, a somatic cell therapy medicinal product, a tissue engineered medicinal product, or a combination product (cell or tissue with an integrated medicinal device). The nonclinical development of cellular therapies classified as ATMPs will form the basis of this chapter.

With approximately 270 tissues in the body [9], there is a broad variety of targets for cell-based therapies. Advances in our understanding of cellular, molecular, bioengineering, and systems biology have all led to an increased exploration of the therapeutic potential of cell-based therapies. Cell therapies may be derived from a patient's own cells (autologous therapies), using cells from a donor (allogeneic therapies), or a combination of cells with scaffolds. Therapies may be based on adult cells or derived from differentiation of an appropriate stem cell progenitor. In some cases, the cells are genetically modified by chemical or genetic means resulting in a product that is both a cell and gene therapy product. Cells can be infused intravenously, directly delivered into damaged tissue, or contained within an encapsulation device or scaffold.

Across a broad spectrum of disease indications, from treatment of injuries to the central nervous system, inflammatory conditions, and ischemic stroke to cancer management, cellular therapies are in development and are being derived from a range of cell types [10] (Figure 1). Even for a given disease indication many different cell sources may be considered. For example, for the replacement of β-cells for the treatment of diabetes [11], options range from the relatively simple islet cells harvested from cadaveric donors for direct transplantation to the recipient (regulated as a tissue product) to therapies in development derived from pluripotent cells [12], pancreatic progenitor cells [13], and the exocrine enriched fraction from the islet production procedure [14], which will be viewed as medicinal products. Each cell type has its own attributes, and taken together with the intended clinical use and the impact of the clinical environment, all influence the development of the nonclinical strategy.

Defining a nonclinical pathway for a cell therapy is, therefore, potentially challenging [15]. The nonclinical program will comprise a series of in vitro and in vivo studies, which will generate the experimental data to support the use of the novel

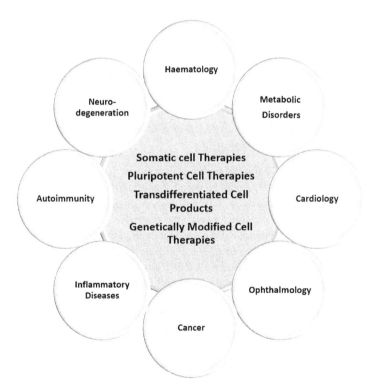

Figure 1 Cell types in therapeutic development for a range of clinical indications: A variety of cell types, including somatic cells (HSC, MSC), pluripotent cell (ESCs, iPSCs), transdifferentiated cells, and genetically modified cells (e.g., T cells) are being explored nonclinically for their potential to treat a variety of diseases.

therapeutic in clinical trial. The studies will support the proof-of-concept of efficacy and provide insight into the potential safety issues, providing the information to assess the risk–benefit ratio for the intended clinical population. The studies will need to meet the recommendations of regulatory guidelines and, where applicable, the requirement for Good Laboratory Practice (GLP). In this chapter, we will provide a perspective on the nonclinical assessment of cell therapy products intended for therapeutic clinical use. The concept of the risk-based approach will be explored, and the importance of a scientifically designed nonclinical program will be discussed.

2. Types of Cell-Based Advanced Therapy Medicinal Products and their Safety Considerations

Cell therapy is one of the most rapidly evolving areas of clinical research with a large number of different cell types being explored as potential therapeutics. The cell choice for a given therapy may be challenging. There is the need to be able to generate sufficient material that remains functional, and this may require the differentiation of cell product. The cells may need to be genetically manipulated and may be allogeneic, raising the risk of immune responses. The source of cells, therefore, influences the potential safety concerns and the design of the safety program. While some concerns are general, others are more specific to particular cell types. Presented below are summary descriptions of cell types that are representative of the starting material for cell therapy products entering clinical trials.

2.1 Somatic Stem Cell Therapies

Somatic stem cells are found naturally within cell niches in differentiated tissue throughout the body. Initially involved in organ growth, somatic stem cells in the adult ultimately have a homeostatic role. By responding to physiological and pathological signals, somatic stem cells can give rise to mature effector cell types through a process of cellular differentiation—effectively repairing tissue damage and replacing lost cells [16]. A wide variety of somatic stem cells are being explored for their potential to repair tissue, and among these, HSCs and mesenchymal stromal cells (MSCs) are the most extensively studied.

HSCs represent the prototype of multipotent adult tissue stem cells and are defined by their capacity to self-renew and to differentiate into all blood cell lineages while retaining robust capacity to regenerate cells of the blood and immune system (hematopoiesis). HSCs are found in cord blood and postnatally in the bone marrow and mobilized peripheral blood. The self-renewing HSCs progress through various intermediate maturational stages generating multipotent progenitor cells, which in turn give rise to oligo-potent progenitors that ultimately produce effector cells—in this case, cells of the hematopoietic system [17]. HSCs are in routine clinical use for the treatment of a variety of blood cell diseases, including leukemic and autoimmune disorders. Both HSC and the resulting progenitor cells are being investigated for their utility as cellular therapeutics for a range of conditions.

Box 1 The International Society of Cellular Therapy (ISCT) Definition of an MSC

- MSCs are considered mesenchymal stromal cells, not stem cells (Horwitz et al., 2005).
- MSCs must be plastic-adherent when maintained in standard culture conditions.
- MSCs express major histocompatibility complex (MHC) class I molecules and surface CD antigens CD105, CD90, CD73, and CD44 and lack expression of hematopoietic markers CD45, CD34, CD11b, and CD19.
- MSCs must also retain the ability to differentiate into osteoblasts, adipocytes, and chondroblasts in vitro [101].

MSC-based products are also being extensively studied as cellular therapies. MSC populations have now been isolated from a variety of tissues, including bone marrow amniotic fluid, adipose tissue, skin, and dental pulp and can differentiate in vitro into a variety of mesenchymal lineages such as osteoblasts, chondrocytes, and adipocytes (reviewed in Ref. [18]). MSCs are classified in accordance with the International Society of Cell Therapy (ISCT) guidelines (Box 1), although some argue that the global definitions of MSCs may be overly simplistic given the range of tissue sources and subpopulations with specific properties. Some cellular therapies termed MSC-like products may be true MSC subpopulations. MSCs are highly metabolically active and differences in metabolic profile may not truly represent a different cell subpopulation but more accurately reflect differences in cell culture conditions and intended clinical application.

A defining factor of MSCs is that, although highly proliferative in culture, they are not immortal. Typically MSC cultures show senescence after 30 to 60 doublings, although if they are re-plated at low density, they will reacquire the characteristics of early progenitor cells [19]. The early hypothesis was that MSCs may engraft to repair tissue; however, in many subsequent animal experiments in a range of disease models, the clinical benefit was observed without significant engraftment. There is now a body of evidence that, for many therapeutic applications, the clinical benefits result from a paracrine action of MSCs with the MSCs homing to and modulating the local microenvironment at sites of tissue injury to promote healing [20]. The most advanced MSC therapy development programs are based on the therapeutic benefit of these immune-modulating effects. The mechanisms underlying tissue regeneration and environment modulation by therapeutic doses of MSCs are the subject of extensive research, particularly the extent to which the two processes may intersect.

There is extensive knowledge of somatic cell therapy biology and potential safety risks, particularly for MSCs, with a low tumorigenicity risk [21] and low immunogenic potential [22,23] reported to date. However, a general assumption of safety cannot be made. For any given application there may be complexities, from the environmental milieu into which the MSCs are delivered, the expected therapeutic mechanism of the therapy in a given disease indication, cell culture conditions, cell characterization and quality control, and any genetic modification of the product. An assessment of safety will need to be made, although the experimental package may be substantively supported by published data.

2.2 Pluripotent Stem Cell Therapies

Pluripotent cells are fundamentally characterized by their capacity for sustained self-renewal and their ability to differentiate into derivatives of all three germ layers (endoderm, mesoderm, and ectoderm [24]). Pluripotency is typically characterized through the ability of the cells to form teratomas, although other specific characteristics, including epigenetic characteristics and gene transcription profiles, must also be fulfilled. Pluripotent stem cells are the natural units of embryonic generation. At the blastocyst stage of the mammalian embryo, the inner cell mass is capable of differentiating into any cell type, and embryonic stem cells (ESC) derived from blastocysts can serve as a source for differentiated cell therapy products.

Murine ESC were the first mammalian pluripotent stem cells that were isolated [25,26]. The resulting cultures contained populations of cells, which grew as colonies, showed extensive capacity for replication, and were pluripotent as demonstrated by their ability to generate chimeras, transgenic mice, and to differentiate in culture into ectodermal, endodermal, and mesodermal derivatives. It was not until 1998 that the first five human embryonic stem cell (hESC) lines were derived from pre-implantation embryos cultured to the blastocyst stage [27]. Since the initial publication, the generation of hESC's has been reproduced in multiple laboratories, and many more hESC lines have been derived and characterized (greater than 300 cell lines are recorded on the National Institutes of Health Website [28] and the UK stem cell bank Website [29]). Good Manufacturing Practice (GMP) cell lines are now available, and protocols for the production of cells that meet the regulatory requirements for clinical application including the absence of animal-derived culture components have been developed [30]. In addition, detailed protocols have emerged for the production of hESC-derived human cell therapy products.

A breakthrough in the production of pluripotent cells came with the publication of the ground-breaking paper by Takahashi and Yamanaka [31] in which they showed that enforced expression of four key transcription factors—Oct4, Sox2, Klf4, and c-Myc—could reprogram mouse nonpluripotent somatic cells such as fibroblasts to pluripotency and achieve similar developmental potential as ESCs. These new cells were termed induced pluripotent stem cells (iPSCs). Although iPSCs share the general properties of pluripotency with ESCs including morphology, pluripotency, self-renewal, and patterns of gene expression, many differences are reported, and our knowledge of the full biological characteristics of reprogramming is incomplete. Differences in reprogramming strategies and culturing protocols may have an influence but may also reflect the extent of epigenetic changes erasing the somatic cell-specific program. Recent publications highlight the possibility that during reprogramming point mutations occur in the DNA that may increase the risk of tumorigenicity of an iPSC therapy [32,33].

Numerous nonclinical animal studies have demonstrated that the differentiated derivatives of ESCs and iPSCs may provide functional replacements for diseased tissues, and clinical trials are currently underway for hESC-based cellular therapies for spinal cord injury [34], macular degeneration [35], and diabetes [36], and the first iPSC-based therapy for macular degeneration initiated in 2014 [37]. The path to initiating these clinical trials has been long. Significant nonclinical packages were required to address the safety concerns of allogeneic human pluripotent stem cell-derived therapies. Studies were required to address the risks of tumorigenicity (teratoma formation from residual undifferentiated cells in the final product), biodistribution, and persistence for products expected to engraft and remain long-term and the risk of an immune response to the administered product [38]. As more therapies are developed and an increasing database of nonclinical studies is produced, a greater understanding will develop on the safety risks of pluripotent stem cell therapies. Key to achieving this, however, will be the publishing and sharing of safety studies.

2.3   Transdifferentiated Cell Therapies

Transdifferentiation differs from the induced pluripotent cell procedure in that transcription factors are used to convert a given cell type directly into another specialized cell type, without first forcing the cells to go back to a pluripotent state. Fibroblasts, for example, can be converted directly into muscle cells at very high efficiencies using the transcription factor MyoD. It was initially thought that conversion occurred only between relatively related cell types (cells from the same germ layer lineage). However, a study by Vierbuchen et al. [39] showed that treatment with a combination of neural transcription factors enabled fibroblasts to be converted to neuronal cells (cells of different germ layer lineage). The first transdifferentiated cell products have yet to enter into clinic trial, but a number of products are in nonclinical development. Currently, many protocols use viral vectors to introduce the required combinations of transcription factors, and this brings with it the risk of insertional mutagenesis. The impact of transdifferentiation on the epigenetic, immunological, and functional characteristics of the derived product will be key to assessing safety.

2.4   Gene-Modified Cell Therapies

There is growing interest in combined gene and cell therapy approaches [40]. These therapies are based on the ability to isolate stem, progenitor, or differentiated cells and genetically modify them to correct genetic mutations or confer altered activity/specificity. Genetic modification of HSCs has been performed, and these therapies are being explored as a therapeutic option for a broad spectrum of genetic and acquired disorders that affect hematopoietic and other tissues [41]. Another application, and one that is receiving significant interest, is adoptive immunotherapy, where the antigen specificity of T cells is modified to harness the power of the immune system by enhancing the T cells ability to selectively target and destroy cancer cells in particular [42]. A further therapeutic option is the use of genetic modification to confer novel functions such as the secretion of gene products with paracrine effects that may have biotherapeutic potential.

Both viral and nonviral methods are being explored for the genetic modification of cells. The major viral systems include retroviral, lentiviral, and adenoviral transduction systems, whereas nonviral methods comprise both chemical (e.g., lipofection) and physical methods (e.g., electroporation). Genetic modification may be stable or transient. In a transient transfection, the introduced gene is only expressed for a few weeks, whereas stable transfection allows the expression of the gene over an extended period of time, potentially permanently. Genetic modification brings with it specific safety risks that need to be considered during nonclinical development, including the potential for oncogenic transformation of the genetically modified cells due to insertional mutagenesis and, when utilizing viral vector systems, the potential for formation of replication-competent viruses. In addition, safety concerns specific to a given class of therapy will need to be considered and assessed.

2.5 Combination Products

Cell-based therapies may require the use of matrices, three-dimensional scaffolds, or specialized devices to aid their delivery or function. The addition of these technologies to a cell therapy product can add to the challenges of the nonclinical program. Classical cell-based tissue engineering products involve the ex vivo cell seeding of a three-dimensional scaffold. The scaffold may be generated from biomaterials or decellularized tissue-derived scaffolds. Recent advances utilize three-dimensional-shaped scaffolds that provide organ shape and bioresorbable substrates for cell growth, which may be recellularized ex vivo, as classically, or alternatively reseeded in situ. Cells may also be delivered on matrices, such as the delivery of iPSC-derived retinal pigmented epithelial (RPE) cells on collagen gel sheets into the eye [43]. Transplantation of RPE in a sheet form is proposed to enable the administered RPE to exert physiological function more effectively than in suspension and may also facilitate the adaption of transplanted cells to the subretinal tissues [43]. Encapsulation is another area in which a cell-based therapy may be combined with a biomaterial. Encapsulation of cells provides the opportunity to minimize the risk of cell distribution, protect the cells from unwanted immunogenicity, and direct delivery of a cell-based product to a specific tissue location [44].

Combination products can be generated from a variety of materials: simple biomaterials such as hyaluronic acid, bone substitutes or alginate-type materials, bioresorbable substrates, smart biomaterials that include thixotropic, thermo-responsive, growth factor-encapsulating or in situ self-assembly properties, and finally, tissue-derived scaffolds (e.g., decellularized organs with the added benefits of native biomechanical strength and matrix factors). For biomaterials as well as decellularized scaffolds, the mechanical properties and degradation kinetics should be adapted to the specific application to ensure the required integrated product functional properties.

Appreciation of the inherent diversities of organ systems, the biomechanical and biophysical constraints on the tissue and scaffold, and the cognate requirements of the cells such as the development of vascularization in a three-dimensional tissue are essential for the development of integrated therapies. Being able to address this in the nonclinical setting can be very challenging, including the choice and relevance of animal models to address aspects of a products safety and challenges of surgical procedures. In addition, new biomaterials will be subject to the requirements of ISO 10993 relating to the biocompatibility of medical devices, and this needs to be considered and factored into the nonclinical program.

3.   Regulations and Nonclinical Studies

A wide range of regulatory guidance exist to help developers understand the nonclinical safety requirements for cell therapy products (Table 1). However, due to cell therapy being a rapidly evolving field with complex and challenging science and therapeutic uses, the development of an appropriate nonclinical program to assess the safety of cell therapy products is often assessed on a case-by-case approach, and as such, early dialogue with regulatory agencies is encouraged. Early engagement enables sponsors to better understand current regulatory concerns and expectations prior to committing to expensive or inappropriate studies and, thereby, ensures appropriate design of the nonclinical package required for clinical trial and licensing. Worldwide, many different national regulatory bodies are responsible for reviewing the nonclinical efficacy and safety data packages that will be used to support the clinical translation of a cell therapy product (Box 2).

Engagement with the regulatory agencies, however, can occur prior to submission of the completed nonclinical data package to support a clinical trial application

Table 1 Key Regulatory Guidelines

Agency	Guidance
EMA	• Guideline on human cell-based medicinal products (EMEA/CHMP/410869/2006)
EMA	• Guideline on the nonclinical studies required before first clinical use of gene therapy medicinal products (EMEA/CHMP/GTWP/125459/2006)
EMA	• Guideline on quality, nonclinical, and clinical aspects of medicinal products containing genetically modified cells (EMA/CAT/GTWP/671639/2008)
EMA	• Reflection paper on stem cell-based medicinal products (EMA/CAT/571134/2009)
EMA	• Guideline on the risk-based approach according to annex I, part IV of Directive 2001/83/EC applied to advanced therapy medicinal products (EMA/CAT/CPWP/686637/2011)
EMA	• Reflection paper on management of clinical risks deriving from insertional mutagenesis (EMA/CAT/190186/2012)
FDA	• Guidance for industry: Preclinical assessment of investigational cellular and gene therapy products (FDA 2013)
FDA	• Briefing document—testing for replication-competent retrovirus (RCR)/lentivirus (RCL) in retroviral and lentiviral vector-based gene therapy products—revisiting current FDA recommendations
FDA	• Guidance for industry: Formal meetings between the FDA and sponsors or applicants (2009)
ICH	• Detection of toxicity to reproduction for medicinal products & toxicity to male fertility S5(r2) (2005 Addendum)
ICH	• Preclinical safety evaluation of biotechnology-derived pharmaceuticals S6(r1) (2011 Addendum)
ICH	• Safety pharmacology studies for human pharmaceuticals S7a (2000)

Box 2 Examples of Worldwide Regulatory Agencies

- The Center for Biologics Evaluation and Research (CBER)/Office of Cellular, Tissue and Gene Therapies (OCTGT) in the United States
- Regional regulatory bodies in Europe such as the Medicines Healthcare Products Regulatory Agency (MHRA) in the United Kingdom and the Paul Erlich Institute (PEI) in Germany
- The regulatory bodies in Australasia include the Korean Food and Drug Administration (KFDA), the Ministry of Health, Labor and Welfare (MHLW), and Pharmaceutical and Medical Devices Agency (PMDA) in Japan, and the Therapeutic Goods Administration (TGA) in Australia.

(CTA)/investigational new drug (IND) application. From a very early stage in the cell therapy product development, a variety of mechanisms exist to allow sponsors to engage with the regulators to seek advice. In the United States, there are two types of advice meeting: the pre-pre-IND meeting and the pre-IND/Type B meeting. The pre-pre-IND meetings are informal meetings specifically designed for the discussion of nonclinical studies. These meetings allow the sponsor the opportunity to seek an early opinion from Center for Biologics Evaluation and Research (CBER) into the proposed nonclinical plans for a given cell therapy. While the discussions are nonbinding, they can help prevent unnecessary studies and identify gaps that will need to be addressed as part of the nonclinical program. The information necessary to engage in these early discussions with the CBER is typically provided in a document no greater than 20 pages in length. This dossier should provide a comprehensive summary of nonclinical data already obtained (including in vitro and cell characterization work) and proposed nonclinical plans plus a brief (approximately one page each) description of the clinical product and the clinical trial design to provide context and background for the therapy and program. The dossier should also include specific questions that the sponsor wishes to ask (Box 3). This dossier should be sent directly to the CBER with a request for a pre-pre-IND consultation. Meetings are approximately five weeks from the date the package is submitted. While no minutes are generated and the recommendations are nonbinding, they are intended to help the sponsors in the preparation of the subsequent pre-IND submission, which are valuable meetings in planning the whole of the product development program (including manufacturing and clinical aspects), especially if the sponsors questions are not fully answered by guidance and other information provided by CBER.

In the United Kingdom, the Medicines Healthcare Products Regulatory Agency (MHRA) offer various advice meetings including scientific advice on the development of cell therapy products. These meetings, similar to the pre-IND meeting, can include seeking advice on the proposed nonclinical testing of a cell therapy product. The MHRA prefers that the questions are prospective and concern the future development of the cell therapy product. A dossier will need to be submitted prior to the meeting, and typically a brief presentation is given on the day to guide the discussion. Similar scientific advice meetings are held by the majority of European Union regulatory agencies.

Box 3 Example Questions to Ask Regulatory Agencies Relating to the Nonclinical Program

- A series of in vitro studies are being performed to examine the ability of the cell therapy to act as targets for the adaptive and innate immune system, to examine their susceptibility to cell-mediated and serum cytotoxicity, and to stimulate T cell proliferation. Does the agency agree the proposed studies are sufficient to address the risk of immunogenicity and support the proposed clinical study?
- We are proposing to conduct a GLP teratoma study in NOD/SCID mice. This study will examine the potential of the pluripotent-derived stem cell therapy to form teratomas following intramuscular injection. Immunohistochemistry assessment will be made of masses to confirm they are of human origin. Does the agency agree that the proposed study is sufficient to assess the risk of teratoma formation?
- We have chosen to conduct our key nonclinical studies in the pig. Use of this model is consistent with the intended clinical use of the integrated product that will be inserted into the kidney using the same surgical procedure to be used clinically. To assess the risks of the novel surgical procedure and general safety of the product, does the agency agree that the proposed study in pigs in combination with the pilot study data is sufficient to support the clinical study?
- In the proposed mouse study, the following tissues will be examined for the presence of human-derived cells by quantitative real-time polymerase chain reaction (qPCR) at three time points (1 day, 1 month, and 6 months) at termination: heart, lungs, liver, kidney, brain, and gonads. Does the agency agree that the proposed study in mice by the clinical route of administration using a validated qPCR assay to detect human cells is sufficient to assess the persistence and biodistribution risk of the cell therapy product?
- Does the agency agree that the efficacy studies conducted in the chemical-induced (streptozotocin) murine model of diabetes support the proposed clinical trial?

To assist organizations that are developing novel medicines such as cell therapy products, the MHRA has recently set up the Innovation Office as part of the UK government's strategy for life sciences. The Innovation Office has been set up specifically to deal with enquiries and questions that are related not only to potentially innovative science but also to the innovative or novel approach to the regulation or manufacture of medicines. There is a further subgroup of the Innovation Office, The One Stop Shop, which is particularly helpful to developers of cell therapy products because it will gather all relevant Competent Authorities in the UK (MHRA, HTA, HFEA, and DEFRA; Box 4) together to give a consolidated response to developers. Enquiries can be submitted to the Innovation Office using an online form [45]. The MHRA aims to provide a response within 20 days, and depending on the nature of the enquiry, the response will consist of either a simple answer or a recommended course of action, which may involve regulatory or scientific advice. Where sponsors have a query regarding a specific aspect of a proposed nonclinical study for an innovative product, such as the proposed study duration, this may be an appropriate route to seek regulatory opinion. If the product is not truly innovative in nature, the MHRA will

---

Box 4 UK One Stop Shop Competent Authorities

- Medicines Healthcare Products Regulatory Agency (MHRA)—Competent Authority responsible for medicines and blood
- Human Tissues Authority (HTA)—Competent Authority responsible for cell and tissues
- Human Fertilisation and Embryo Authority (HFEA)—Competent Authority responsible for research and product development associated with human embryos
- Department for Environment, Food and Rural Affairs (DEFRA)—Competent Authority responsible for assessment of environmental risk associated with the use of gene therapy products

---

divert to the appropriate advice forum, which may either be an advice meeting solely for preclinical advice or may also include clinical, quality, and regulatory advice.

Output from scientific advice meetings is highly valuable and highly recommended. During regulatory engagement, the sponsors should be prepared to discuss the proposed pivotal study designs, to address the choice of animal model, number of animals selected, and study duration, to discuss the use of potential in vitro study alternatives to in vivo studies, and to identify risks that they may not be able to address during the nonclinical program and the potential mitigation that can be applied in the clinical setting. The regulatory engagement allows questions to be raised if there are significant concerns with the proposed approach. It is the responsibility of the sponsor to present to the regulatory agency a proposed plan that will be the subject of discussion. The regulatory agencies will not design the nonclinical plan for a program. This may seem intimidating to scientists who may not have been previously involved in a regulatory engagement and people are concerned that they may recommend the "wrong experiment." The role of the engagement is to allow the opportunity for discussion. The regulatory agencies, through their experience gained from other programs in development, may be able to highlight weaknesses in a program and other areas that need to be addressed. In addition, they may recommend amendments to the proposed study durations or advise additional study endpoints, particularly with respect to genetically modified cell products. Moreover, this is the opportunity to discuss the choice of animal model(s). The advice can prevent unnecessary nonclinical studies being performed, provide confidence in the acceptability of the proposed plans including animal models, and identify areas where additional information may be required or where the agency has specific regulatory concerns; this should help prevent delays to the subsequent CTAs.

4. Nonclinical Assessment—The Risk-Based Approach

Unlike the nonclinical pathway for small molecule and biologic-based therapies, the requirements for a cellular therapy product can at first seem unclear, and many researchers are unfamiliar with the regulatory recommendations and requirements.

The basic aim for any nonclinical program is to determine the efficacy and safety of the product; however, with a cellular therapy, a key challenge can be determining how to do this. Although cellular therapies can share some of the same principal characteristics, it is recognized by the regulatory agencies that cellular therapies are not a homogeneous class of products [46]. In addition, the level of scientific knowledge and clinical experience of a given cellular therapy is highly variable. For example, extensive clinical experience has been obtained with MSC-based therapies compared with the limited clinical experience of pluripotent cell-based therapies. It is also important to work closely with the scientists developing the cell production and manufacturing processes as these can directly impact the biological characteristics of the cellular therapy and potentially the products safety profile. Given the product-specific attributes of most cellular therapies, a case-by-case, risk-based approach can be taken when designing the nonclinical testing programs.

The risk-based approach [47] is based on a series of generic scientific questions that could apply to any cell therapy product (Table 2). The risk factors are related to the quality, biological activity, and clinical application of the cell therapy. By determining the risk for a given therapy, the extent of the nonclinical package can be determined. The factors associated with a given risk are product specific and, in many cases, multifactorial, and each of these needs to be considered as part of the overall assessment. Once risks have been identified, it is then possible to determine the process for collecting data to assess each of them; this may be a combination of in vitro and/or in vivo studies but may also constitute a paper-based exercise whereby applying knowledge from the literature, a scientific argument can be formulated to address a specific concern. The risk-based assessment is a living document, and changes to the risk profile of a given cell therapy can become apparent as product development proceeds. For example, from study data, changes to manufacturing processes that alter the characteristics of the cell therapy or through information in the community on related products. By starting the risk-based process early in the development phase of a product and maintaining it as the product matures, potential gaps in knowledge can be identified and subsequently addressed, minimizing delays to the development program.

There needs to be a balance between the potential risk, the ability to assess that risk, and the results of analytical assessment of the product characteristics, as well as proof-of-concept animal studies and an understanding of the clinical condition [48] (Figure 2). Unlike small molecule- and biologic-based therapies, the characteristics of a cell therapy may, and indeed very likely will, change upon administration. Cells may be delivered into highly inflammatory environments that can alter the expression of cell surface molecules, the patient may be taking multiple medications that could affect cellular function, or the patient could generate an anticellular therapy immune response; all of these factors can alter the risk profile of a product and need to be considered as part of the risk-based assessment. It is also important to consider past and ongoing regulatory examples. Extensive regulatory knowledge has, for example, been gained for cellular therapies based on immune-modulation by MSCs, and specific aspects such as the risk of tumorigenicity may be better understood and the nonclinical requirements better established, and this knowledge can be applied to developing the nonclinical requirements [21]. Based on all of these aspects, it should be possible to

Table 2 The Risk-Based Approach to Aid the Design of the Nonclinical Program

Risk	Risk Factors May Include:	Mitigation May Include:
Relevance of animal model	• Disease model has different pathophysiology and is not reflecting human disease, e.g., acute versus chronic condition • Altered sensitivity of model to cell therapy product • Model is not predictive of immunogenicity in patients • Age, dosing, immunocompetence, and duration of available animal study may not be predictive for the risk of tumor formation • Model does not permit use of planned clinical delivery/procedure • Dose level(s) that can be administered especially to a murine model; relevance to the clinical setting	• Model is the current gold standard • Tiered approach to model selection including the conducting of pilot studies to confirm relevance of test species • Multiple animal models are used to adequately identify functional aspects and potential toxicities of the cell therapy under investigation • Nonstandard models such as genetically modified rodents or large animal models may be relevant with adequate justification • Route of administration/delivery procedure could be assessed in a larger animal model • Nonclinical in vitro assays to assess aspects of the biological activity of the cell therapy (e.g., immunological response profile) to provide supporting proof-of-concept information • Justification of dose level with supporting data is provided for the specific dose levels selected
Heterogeneity or insufficiently defined cell product	• Undifferentiated and/or undesirable cells including pluripotent cells • Presence of cells with inappropriate characteristics • Potential for immune reactions against activated autologous cells	• Each lot of investigational cell therapy product used in the nonclinical program is characterized according to appropriate criteria, consistent with the stage of product development • Similarities and differences between cell therapy lots intended for nonclinical use and lots intended for clinical use will be highlighted and discussed in the regulatory submission • Report isolation efficiency and viability for each nonclinical study • Confirm lack of specific contaminating cells, e.g., pluripotent cells • Pilot nonclinical studies run with activated autologous cells to understand risks of immunogenicity

immunogenicity	• Possible HLA mismatching • Proinflammatory profile of disease model • Distribution of cells may increase risk of immune rejection • Extent of immune privilege of site of administration • Repeat dosing in the clinical setting	…*cy is supported by a comprehensive* package of in vitro studies • Literature review to support immunological status of cells • Literature review to support extent of immune privilege at the site of administration. • Encapsulation of the cell therapy • Use of species-specific autologous cells to assess specific immunological aspects; including scientific justification for approach
Tumor formation	• Risk of cell transformations due to culture conditions • Risk of genetic stability due to long-term culturing • Tumorigenic potential of contaminating undifferentiated pluripotent cells • Tumorigenic potential affected by site of administration	• Cytogenetic and genetic characterization of the final clinical product • Product is low passage • Cell characterization assays • Soft agar or alternative in vitro assays • Prospectively designed study, of sufficient duration, in immunocompromised mice to assess the risk of tumorigenicity • Perform a combined tumorigenicity and biodistribution study • In-life monitoring of cell distribution combined with comprehensive tissue list taken at necropsy for qPCR and/or IHC analysis • Scientific rationale based on previously published experience, including information on product characteristics to support comparability and relevance of data • Use of species-specific autologous cells to assess specific immunological aspects; including scientific justification for approach
Biodistribution and persistence	• Risk of tumor formation in different organs • Risk of unwanted tissue formation including structural and functional tissue integration • Potential increase in risk of immunogenicity at nonimmune-privileged sites	
Toxicity	• The biological responsiveness of the animal species to the investigational cell therapy product • The mode of action of the cell therapy, e.g., risk of toxicity due to secretion of bioactive substances • Risk of cell overgrowth • Risk of ectopic tissue formation due to cell biodistribution • The pathophysiology of the animal disease model	• Published nonclinical and clinical safety information on similar products highlighting known toxicities or adverse events • Previous nonclinical/clinical experience with the proposed clinical delivery device/delivery procedure • For the toxicology assessment; use of an animal species in which the cell therapy product is biologically active; supporting data should be provided that justify species selection

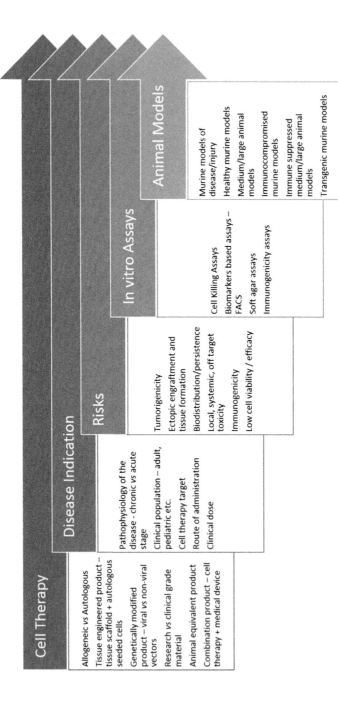

Figure 2 Design considerations for the development of safety programs for investigative cell therapies: Multiple factors influence the final design of the nonclinical safety program for a given cell therapy. These will be influenced by the cell therapy product and expected persistence, available animal models, final clinical program design including proposed clinical dose, and dosing route. There is no one-size-fits-all design, and a scientifically justified program will be required.

develop a strong scientific rationale for the proposed nonclinical package. It is highly recommended that the sponsors engage regularly and early with the regulatory agencies, to confirm that the proposed nonclinical plans for a given cellular therapy are acceptable to the regulatory authorities (see Section 3). It will also be important to discuss whether or not the team intend to run studies to GLP.

5. The Requirement for Good Laboratory Practice

GLP is a set of principles that provide a framework for how pivotal nonclinical studies are planned, performed, monitored, recorded, and reported. GLP was established to promote the quality and validity of test data used for determining the safety of medicinal products. GLP is an official regulation that was established in the United States in 1978, following an in-depth review of practices at toxicology laboratories. The review revealed poor laboratory practices, the inaccurate and incorrect recording of data, and the inability to reconstruct completed studies that had subsequently been submitted to the FDA as part of the assessment of pharmaceutical safety. As a result, the Organization for Economic Co-operation and Development (OECD) established an expert working group and, based on the US regulations, established the OECD Principles of GLP. The Principles of GLP were officially recommended for use in member countries by the OECD Council in 1981. One of the key parts of the regulation is that safety data collected in accordance with the OECD test guidelines and OECD Principles of GLP are accepted in other OECD member countries although how the regulations are implanted may differ (Box 5).

Box 5 Examples of Regional GLP Implementation Differences

Although the Principles of GLP are based on the OECD guidelines, how the regulations are implemented differ between countries.

- In the United States, GLP is governed by the Code of Federal Regulations 21CFR58.
- In Japan, under the provisions of the Pharmaceutical Affairs Act, the Ministerial Ordinance on GLP for Nonclinical Safety Studies of Drugs have been established.
- In the United Kingdom, compliance with the Principles of GLP is a legal requirement (Statutory Instrument 1999 No. 3106: The Good Laboratory Practice Regulations; Statutory Instrument 2004 No. 994: The Good Laboratory Practice [Codification Amendments Etc.] Regulations) for test facilities that undertake such studies, and a laboratory has to belong to the UK GLP compliance monitoring program, run by the UK GLP Monitoring Authority (UK GLPMA).

When running studies on GLP, it is essential to understand the local requirements to ensure that GLP compliance is met and false claims are not made.

For all medicinal products entering clinical trials, pivotal safety data is required to perform GLP. This general expectation also applies to cell-based therapies. In both the US and EU cell and gene therapy guidelines, the general principle of compliance with GLP regulations remains (Table 1). However, and importantly, the regulatory guidances also acknowledge that some or even all of the toxicology assessments of cell-based therapies may not be able to fully comply with the GLP regulations. Key safety studies, for example, may only be possible in animal models of disease or injury. In many cases, these are established in academic or specialist research facilities and may not readily be transferred to a GLP facility. In addition, specialist assays, for example assessing cell fate or specific immunological endpoints, may not be available at a GLP facility but clearly could provide valuable data to help assess the safety of a given cell product.

In vitro and in vivo pharmacology and proof-of-concept efficacy studies are not required to be performed to GLP. For a cell therapy product, this includes studies to assess the biodistribution and persistence. However, in many cases during the development of a cellular therapy, the delineation between an efficacy study and safety study is not clear. Test material may be limited, and nonclinical proof-of-concept studies may be designed to collect multiple endpoints including key safety endpoints. When developing the program of studies to assess the safety of a cell therapy product, the requirement for those studies to be GLP must be considered.

Where the in-life phase of a pivotal safety study cannot be performed at GLP facility, safety endpoints such as clinical pathology and histology could potentially be outsourced to a specialist GLP facility, and consideration of this approach is highly recommended. Another option is to perform pivotal safety studies in collaboration with a nonclinical contract research organization (CRO). If specialist surgical procedures are required to administer a cell therapy, it may be possible for key surgical staff and equipment to be brought into a GLP facility. Alternatively, the surgical procedures may be performed at the specialist facility, and following a designated period of recovery, the animals can be transferred to the nonclinical CRO for the remainder of the study. Early interactions with CROs are required to discuss the feasibility of these options. In our experience, such collaborative approaches, while not fully GLP compliant, utilize the skills of the different partners and can deliver high-quality nonclinical data to support the translation of the product into the clinic.

When it is not possible to work with a CRO and the safety study cannot be performed to GLP but includes key safety endpoints, a prospectively designed study protocol is required (Table 3). Ideally, a study lead or study director will be appointed to oversee the performance of the study. The study lead will ensure that the protocol is provided to all personnel who will be taking part in the study, that the procedures in the protocol are followed with any amendments or deviations (see glossary) documented and their impact assessed, that all the raw data is documented and recorded, and at the end of a study, a study report is generated and all study documentation is archived. By performing a study in this manner, a research laboratory should be able to provide confidence in the quality and validity of the pivotal safety data presented as part of regulatory submissions. Where pivotal safety studies or components thereof cannot be performed in compliance with GLP, a brief statement for the reason of noncompliance must be provided with each study and clearly stated in regulatory submissions.

Table 3 **Basic Contents of a Prospectively Designed Study Protocol**

Protocol Section	Contents
Study identification	Descriptive title including identity of test item (cell therapy)
	Identity of study lead
	Identity of the sponsoring organization (e.g., academic group; company)
	Unique study identifier
	Name and address of study location
	Dates of study (start and proposed end date)
Signature page	Study lead signatures and any relevant other facility signatures
Table of contents	Include all sections including appendices
Protocol distribution list	All staff who will be working on the study need to be identified and their role. All need to receive or have access to a copy of the protocol
Calendar of events	Table detailing which events will happen to which animal or group of animals on which planned days (+/− x days where relevant)
Introduction	Brief background
Objectives	A statement that details the nature and purpose of the study
Records to be maintained	All other records that would be required to reconstruct the study and demonstrate adherence to the protocol; including but not limited to protocol, final report, histology and pathology records, standard operating procedures, study-related correspondence, amendments, and deviations
Humane endpoints	Detail proposed humane endpoints for study
Test system	Species and number of animals and justification for species choice and method of administration
Husbandry	Include information on how the animals will be accommodated and their diet and drinking regimen
Experimental procedures	Detail experimental plan, where local standard procedures are used either reference and maintain the record or provide details of method in the protocol.
In-life procedures	Include in vivo imaging plans (where relevant), body weights (how and when they will be measured), general health observations, blood sampling for immunology, or other assessments
Termination and necropsy	Include details of planned procedure including list of all tissues that will be collected for histology or PCR analysis
Appendices	Include forms you will use to record study data, etc., not standard laboratory forms

When planning the pivotal nonclinical program that contains studies that cannot be performed to GLP, it is highly recommended to engage early with the regulatory agencies as to the acceptability of the approach. The acceptance of the approach, however, may be different for different regulatory agencies. If in the clinical development program the plan is to run the initial clinical trial in multiple countries, the acceptance of the approach to the nonclinical program by the different agencies needs to be considered. This can also be an issue if the cellular product in development is taken to different countries later in the clinical development process. The greater the documented evidence you have to support the quality and validity of the studies performed, the more likely you are to mitigate risks later in the product's development.

Some laboratories, for example preclinical assay laboratories, while not GLP compliant, claim compliance with ISO standards, in particular ISO 9001. ISO 9001 is a standard relating to the requirements of quality management systems and outlines ways to achieve, as well as benchmark, consistent performance and service. Being in possession of an ISO 9001 accreditation indicates that an organization has a commitment to quality but does not directly assess the quality of a service. To gain laboratory accreditation, laboratories must comply with one of two relevant standards: ISO/IEC 17025 or ISO 15189:2012. ISO/IEC 17025 is the standard that assesses the general requirements for the competence of testing and/or calibration laboratories whose testing is performed using standard methods, nonstandard methods, and laboratory-developed methods. ISO 15189 relates specifically to medical laboratory requirements for competence and quality. A laboratory's fulfilment of the requirements of either of these standards means the laboratory meets both the technical competence requirements and management system requirements that are necessary for it to consistently deliver technically valid test results. Although there are commonalities with ISO 9001, ISO/IEC 17025 and ISO 15189 are more specific in the requirement for competence. Technical requirements include factors that determine the correctness and reliability of the tests performed in laboratory. Both these standards are an accreditation and not a certification of laboratories, but nonetheless, regulatory authorities may use it in recognizing the competence of laboratories that have performed aspects of a nonclinical safety assessment.

6.   General Study Design Considerations

Nonclinical studies are conducted to provide scientific insight into the potential efficacy and safety that may result from the administration of a cellular therapy into humans. The studies are critical to assess the risk–benefit ratio, but a study should also be relevant. A failure from insufficient safety assessment could have very significant clinical consequences for patients resulting in, for example, tumor formation, life-threatening graft rejection, or patient death due to unexpected side effects and could result in a loss of confidence in this class of therapies. However, safety assessments are not absolute. The nonclinical studies will only ever be predictive. The species differences and the challenges in assessing a human cell product in an animal system means that careful consideration of all relevant information will be key to generating the overall risk profile of the cell therapy product. A review of information published on the European Public Assessment Records (EPAR) and the US biologic license application approvals (Table 4) indicates the product-specific nature and range of nonclinical development packages that have been used to support the registration of cell therapy products.

Rigorous criteria need to be applied before a cell therapy product can progress into the clinic. Using the risk-based approach (see Section 4), a comprehensive nonclinical program can be developed to identify and understand potential toxicities before

Table 4 Nonclinical Safety Packages for EMA-approved Cell-based ATMPs and US-approved Investigational Cell Therapies

Product (Approval Year)	Description	Company	Therapeutic Area	Type	Nonclinical Pharmacology/Toxicology Studies
EMA-approved cell-based ATMPs					
ChondroCelect (2009)	Characterized viable autologous cartilage cells expanded ex vivo expressing specific marker proteins	TiGenix NV	Cartilage defects	Autologous	*In vivo* combined pharmacodynamic/pharmacokinetic (distribution)/toxicological studies (non-GLP) in the ectopic mouse (nu/nu model) and in orthotopic models (sheep and goats) EPAR: EMEA/724428/2009
Holoclar[a]	Ex vivo expanded autologous human corneal epithelial cells containing stem cells	Chiesi Farmaceutici S.p.A. Holostem Terapie Avanzate (manufacturer)	Limbal stem cell deficiency (LSCD) due to ocular burns	Autologous	Nonclinical safety studies were not required due to extensive clinical experience (200 patients) gained prior to enforcement of Regulation 1394/2007 for ATMPs when the product was considered a tissue transplant.[c]
MACI (2013)	Matrix-applied characterized autologous cultured chondrocytes	Aastrom Biosciences DK ApS	Fractures, cartilage defects	Autologous	*In vivo* primary pharmacodynamic studies (rabbit, sheep and horse models), in vitro secondary pharmacodynamics studies (cell viability and cell–membrane interactions assessment), long-term safety equine study (pharmacology, biodistribution, and local tolerance), single-dose toxicity studies, local tolerance studies (rat, rabbit and horse) Single-dose toxicity studies in mice (collagen membrane extracts) and in horse (MACI implant). Genotoxicity, local tolerance, cytotoxicity, and sensitization studies (GLP) for the porcine-derived collagenous membrane only (ACI-Maix™) EPAR: EMEA/H/C/002522

Continued

Table 4 Nonclinical Safety Packages for EMA-approved Cell-based ATMPs and US-approved Investigational Cell Therapies—cont'd

Product (Approval Year)	Description	Company	Therapeutic Area	Type	Nonclinical Pharmacology/Toxicology Studies
Provenge (sipuleucel-T) (2013)	Autologous peripheral blood mononuclear cells activated with prostatic acid phosphatase granulocyte-macrophage colony-stimulating factor (sipuleucel-T)	Dendreon UK LtdPharmaCell (manufacturer)	Prostatic cancer	Autologous	Nonclinical safety data obtained from in vivo proof-of-concept studies (rat and mice); conventional nonclinical safety pharmacology studies[b] not deemed necessary due to sufficient clinical safety data EPAR: EMEA/H/C/002513
FDA-approved cell therapy products					
Carticel (2007)	Autologous cultured chondrocytes	Genzyme BioSurgery	Cartilage defects repair	Autologous	*In vivo* safety and efficacy studies in dog and rabbit models; post-approval studies in goat and horse models BLA: STN#103661
GINTUIT (2012)	Allogeneic cultured keratinocytes and fibroblasts in bovine collagen	Organogenesis Incorporated	Oral mucogingival conditions	Allogeneic	Product manufactured in the same way as Apligraf, which was approved as a medical device in 1998 and was intended for the treatment of skin ulcers. Retrospective data obtained from nonclinical in vivo pharmacology and biocompatibility studies (nude mice and rabbits) for Apligraf. *In vivo* studies conducted in nude mice to demonstrate GINTUIT compatibility with periodontal products. Toxicology studies as per ICH guidelines[b] not conducted due to the nature of the product and extensive clinical experience with Apligraf. BLA: STN#1254000/0

| Laviv (azficel-T) (2011) | Autologous cultured fibroblasts | Fibrocell Technologies | Nasolabial fold wrinkles | Autologous | No preclinical safety studies conducted due to lack of appropriate animal models. Safety assessment was based on clinical experience with an equivalent product (Isolagen Therapy[TM]), which was marketed as a cosmetic product prior to it being evaluated under a BLA as a somatic cell therapy product. BLA: STN#125348/0 |
| Provenge (sipuleucel-T) (2010) | Autologous cellular immunotherapy | Dendreon Corporation | Prostatic cancer | Autologous | Nonclinical safety data obtained from in vivo pharmacology and toxicology studies[b] (in mice and rat using rodent product equivalent). Conventional nonclinical safety pharmacology studies were not deemed necessary due to the autologous nature of the product and the patient population of focus evaluated in the BLA. BLA: STN#125197.000 |

Abbreviations: ATMP = advanced therapy medicinal products; BLA = Biologic License Application; CHMP = Committee for Medicinal Products for Human Use; EPAR = European Public Assessment Reports; GLP = Good Laboratory Practices; HPC = hematopoietic progenitor cells; ICH = International Conference on Harmonization.
[a]Recommended for conditional marketing authorization in EU by the CHMP (19/12/2014).
[b]As defined in the ICH guidelines M3 (R2) (EMA/CPMP/ICH/286/1995).
[c]Chiesi—personal communication.

a cell-based therapy enters into clinical trials. Although both tumorigenicity and distribution/persistence to nontarget locations are often cited as specific theoretical concerns for cellular therapies and must be addressed, the routine goals of toxicology testing cannot be ignored, and the principles of nonclinical safety evaluation are the same as for all biopharmaceuticals.

The primary objectives of the discovery phase and the proof-of-concept studies are to establish the scientific rationale and feasibility of the proposed clinical trial, and key to this is an understanding of the clinical population and the pathophysiology of the disease indication. Where animal studies are appropriate, it is important to identify relevant animal models, as will be discussed in detail later (Section 6.2). Where relevant animal models do not exist, alternative strategies need to be considered. An unrelated disease model that has significant commonality of disease pathology may allow specific aspects of safety to be assessed. Alternatively, in vitro models may be more relevant. The antigenic target of a genetically modified T cell therapy, for example, may not have a similar expression pattern on normal cells in an animal model compared to the human. The use of in vitro peptide scanning technologies could be employed to screen for human targets of cross-reactivity. Potential cellular targets can then be assessed as part of an in vitro reactivity study [49].

Early efficacy studies will normally be run with research-grade test cell therapy material. Developing the clinical manufacturing process typically occurs following proof-of-concept studies with the aim to run pivotal nonclinical studies using material made with the final clinical manufacturing process. Important data on safety and efficacy, however, may already be available from completed studies. By recording how test material was manufactured for each nonclinical study and understanding how changes to manufacturing may affect the characteristics of a test material such as effects on cellular morphology, function, and behavior, the full context of the early study data can be presented. This, however, is accepted as more challenging than for a traditional biopharmaceutical as all relevant cellular characteristics for cell therapy function may not be known. One option may be to exploit microarray technologies to examine changes to gene expression patterns, and while these methods can provide valuable information, there can be difficulty in determining what a relevant or critical change is. The cell is clearly multifunctional, and focus needs to be on the primary characteristics that are critical for therapy function and those characteristics that would not be acceptable from a safety perspective, for example, the upregulation of MHC Class II antigens and co-stimulatory molecules in nonantigen presenting cells.

During much of the nonclinical program, products are made at laboratory bench scale with sufficient material made for a given study. The optimal conditions for sample storage are unlikely to be known and variability in product, as well as inherent variability in animal models or in vitro assay variability, can make interpreting results between studies difficult. To understand how one study relates to another, it is important to record basic information about the cell therapy including dose, viability, and the percentage of cells expressing a relevant product characteristic (e.g., the expression of a specific cell surface marker). The cell dose can be expressed in a number of ways depending on the therapy. It can be expressed as the total number of administered cells, the number of cells expressing a specific characteristic (e.g., genetically modified), or the number of viable cells or potential combinations thereof. To define a cell dose by a relevant product characteristic, it is important to have a standardized assay. The assay method should be captured in either a laboratory standard operating procedure or laboratory manual to ensure that the assay is run reliably. Another way to ensure some consistency across studies is to ensure that for each study the test cell therapies have a minimum cell viability—typically ≥70% viability pre- and postdosing to ensure that all animals receive an effective dose and to provide confidence in meaningful study data. Clearly defining the cell therapy dose can be more challenging for an integrated product, and other criteria also need to be considered to provide confidence in how study materials compare.

Data from the nonclinical program is traditionally used to help inform the doses of therapies to be used in a subsequent clinical trial. However, extrapolating an effective cell therapy dose can be very challenging from nonclinical studies. A variety of factors can impact the ability to extrapolate the dose, and these include the clinical route of administration, the biodistribution profile and cell expansion/replication profile, the cell therapy product species specificity, and any immune response to the administered product. Typically, small molecule pharmaceuticals use allometric scaling (e.g., milligrams of drug/kilogram of body weight), but it is not always clear whether this is relevant for a cell therapy product. A number of different methods have been employed to assist clinical dose selection from nonclinical studies. These include dose extrapolation based on body weight, body surface area, or target organ volume—the final choice being scientifically justified.

The regulatory guidances suggest that where ever practical, up to three cell dose levels should be assessed nonclinically, including a minimally effective dose and maximum feasible dose. Caution, however, needs to be employed. The maximum tolerated dose in mice of intravenously administered MSC therapies is typically between 0.5 and 2 million cells. This is to prevent death by lung embolization due to the larger size of the human MSC relative to the mice microcapillaries [50]. This is significantly below safely administered clinical doses, even based on allometric scaling. Another caveat is that some cell therapies can and do increase in number following administration, and adoptive T cell therapies are a classic example of this [51]. T cell therapies are typically administered as a defined number of cells per kilogram of body weight. However, because T cells can replicate and expand after transfer in response to target antigen, the administered cell dose does not resemble the final steady-state number of cells. Cell number will vary among patients as the level of T cell expansion will be patient specific and may not be replicated in the nonclinical setting. It is important to consider how product characteristics and animal model characteristics may impact dose extrapolation.

When cell products are administered directly into a tissue, a more relevant dose extrapolation calculation may be based on tissue volume. For the majority of integrated technologies, it may not be feasible to assess multiple dose levels, and ideally, the proposed clinical dose will be assessed nonclinically, and this may require the use of larger animal models. The number of large animal models of disease is more limited, and immunocompromised models do not exist. It will be important to determine what information the studies will meaningfully provide and whether the studies are relevant. The impact of any immunosuppression on the functionality of the cell product and long-term animal health need to be considered, which will help to determine a meaningful study duration. Where the clinical dose cannot be informed by the nonclinical program, previously published clinical experience with similar or related cell therapy products may provide additional information for estimating a clinical dose. This is entirely reasonable, particularly where there is extensive literature on a given cell therapy type, and again this justification, including scientific relevance, should be included in any regulatory submissions.

Where innovative delivery technologies are to be employed, it is critical to confirm that the procedures are safe. As indicated above (see Section 2.5), any device that will be permanently implanted must meet the ISO 10993 standards. A novel delivery system designed to deliver cells to a given target organ may be difficult to translate to use in the nonclinical setting, and studies to examine the safety of the approach may need to be performed in large animal models such as pigs where relevant disease models may not be available. Where a specific murine disease model is required, different forms of delivery may need to be employed to support efficacy. Justification for the mode of delivery used in nonclinical studies will need to be given and any limitations understood.

The timing of cell therapy delivery relative to the onset of disease will also need to be established. This can be particularly challenging to replicate in the nonclinical setting. Many diseases transition happens from an acute stage to a chronic disease with very different pathophysiological characteristics. If the clinical disease indication is a chronic condition, but the available animal models reflect only the acute condition, the relevance of the animal model may be questionable. The environment in which the cell therapy is delivered may be very different, and chronic conditions may alter the feasibility of novel surgery techniques. Careful consideration needs to be given to the relevance of any animal models of disease, and this will be discussed in more detail in Section 6.2.

6.1   Clinical Product versus a Species-Specific Analog

One question that is often raised is whether the nonclinical program should be based on the clinical product, typically a human cell therapy product or a species-specific analogous cell therapy product. In certain circumstances, the use of a species-specific analog is not appropriate. Tumorigenicity assessment should always be evaluated with the clinical product to ensure the direct relevance of the data. For other nonclinical studies, however, there is no definitive answer, and as is often the case, it depends on the therapy that is being developed.

There are limitations with both approaches, human cells will often need to be administered to an immune-suppressed or immune-incompetent host, and this will limit the ability to assess immune interactions and the impact of the inflammatory environment on the efficacy of a cell product. However, in the absence of immune suppression, the cellular therapy may not persist due to the induction of a xenogeneic immune response against the administered human cell therapy, potentially limiting the useful nonclinical assessment. There may also be intrinsic differences in the properties of the cell in the specific host environment, and limited information may be available to accurately assess the impact of this in the nonclinical setting.

An alternative approach is to study an animal-derived analog to the human cellular therapy. This option may be potentially acceptable for some nonclinical programs, but it requires extensive understanding of the comparability of the animal-derived analog with the clinical product before initiation of pivotal nonclinical studies [46]. The animal-derived analog will require extensive characterization, and depending on the stage of product development, the ability to determine this may be challenging. Animal-derived analogs isolated at a similar cell development stage may have different phenotypic and functional properties, and this can include the excretion of different growth factors and cytokines or species differences in molecular pathways, and this may result in significant biological differences compared to the human product [52]. In addition, the interaction of the animal-derived analog with the disease model microenvironment may differ from the clinical situation, making the translation of nonclinical study findings challenging.

The animal-derived analog also need to be developed according to the same standards as the intended clinical product. However, the isolation, expansion, and culturing methods may be different, and this can result in different impurities and contaminants being present that may alter the risk profile of the analogous product. The cell growth kinetics may be altered, in particular with differences in achieving senescence and chromosome stability. Where the product involves the seeding and growing of cells on scaffold, there may be substantive changes in culturing time and the growth factors required. It is likely, therefore, that additional animal studies will be required prior to initiation of pivotal nonclinical studies. A scientific rationale will be required for either approach, but data from an analogous product to support a clinical trial will require in-depth comparison between the animal and the human cells, particularly if extrapolating a safe starting clinical dose. Despite the challenges, the FDA recently reported [46] that 25% of analyzed submissions (163 products) included nonclinical studies that used an analogous product.

6.2 Animal Models

A critical step in the design of nonclinical in vivo studies for a cellular therapy relates to the selection of an appropriate animal model. The rationale for using animal models in preclinical testing is that animals are predictive, at least to a certain extent, of the human responses and can, therefore, provide a platform to evaluate safety, feasibility, and efficacy of novel therapies before moving into the clinic. Given the heterogeneous nature of cellular therapies with respect to cell source and biological activity, methods of delivery (i.e., use of scaffold or delivery devices), and intended clinical indications, the type and number of nonclinical safety studies and, consequently, the relevant animal models to be used needs to be determined on a product-specific basis and should be supported by a clear, science-driven rationale. The risk-based approach (Section 4) assessing the intrinsic properties of the cellular therapy under investigation can be used to define the critical efficacy and safety aspects to be examined, and this along with the intended clinical indication will help direct the selection of the animal model.

Animal models of disease and injury are largely employed for proof-of-concept studies aiming to evaluate the efficacy and feasibility of a cellular therapy and include induced models (e.g., chemical induction), spontaneous models (genetic variants, which mimic the human condition), and transgenic models (e.g., knockout animal models). Each of these animal models can be further categorized based on the extent to which they model the human condition. Models may be defined as having high fidelity and have a highly relevant biological closeness to the human condition. Isomorphic animals have similar symptoms or anatomy to the clinical condition, but the etiology or the genetic character of the disease may be different. Alternatively, animals may have partial identity in that they may not mimic the entire disease, but there are sufficient similarities that specific, defined aspects of the disease can be examined. A clear understanding of the disease/injury model is, therefore, paramount to appropriately design and ultimately assess study findings.

The majority of studies are performed in murine models due to wide range of models available and the wide range of assay reagents, allowing ease of assessing study endpoints. A central debate, however, is whether studies in large animal models, including pigs, sheep, dogs, and nonhuman primates (NHP), should be considered more relevant to humans and, hence, preferred over studies in small species. The requirement for a large animal study, however, is not standard but must be judged on a case-by-case basis depending on the cellular therapy under investigation (Figure 3). For instance, safety and feasibility assessment of cellular therapies combined to bioengineered natural or

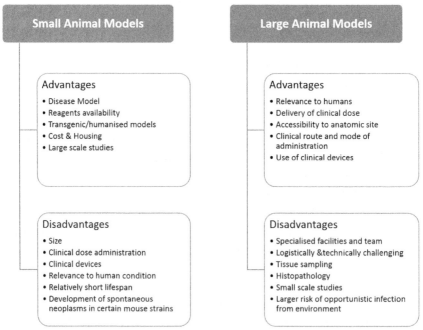

Figure 3 Comparison of small versus large animal models: A variety of factors will influence animal model selection including cell source and biological activity, methods of delivery (i.e., use of scaffold or delivery devices), and intended clinical indications. The major advantages and disadvantages of small and large animal models are detailed here.

synthetic scaffolds for tissue/organ replacement may preclude a priori the use of a small animal species, in which case, a large animal model may be the only viable option. The use of large species can offer some clear advantages with respect to human relevance including analogous physiological and biochemical responses, accessibility to the anatomic site for target product delivery using the intended clinical delivery route and method, and comparable organ/tissue size and morphology allowing assessment of the absolute clinical dose of the therapy and relative pre- and postsurgical treatments. On the other hand, studies on large species remain logistically and technically challenging, with a limited number of animals typically included in the study. In addition, the number of large animal disease models is limited, and in general, the study of the clinical product requires the long-term use of immune suppression.

The majority of in vivo assessments of cell therapy products requires the use of immunocompromised or immune-suppressed animals to prevent the human cell therapy rejection. The impact of immunosuppressive regimens or immune deficiency status of the model will need to be understood both in terms of the health of the animal and any potential impact on the activity and safety profile of the cellular therapy. Immune-deficient animals differ with respect to which components of the immune system are missing, their life expectancy, and incidence of spontaneous lesions (Table 5); a full understanding of these will be required to interpret study data (see Section 7.3). Life expectancy can be limited further if the animal is also an animal model of disease. This may limit the study duration, and careful consideration needs to be given to defining clear study endpoints such as indicators of efficacy, as large numbers of animals may be required to give statistically meaningful information. An alternative option is immunosuppression. A common immunosuppressant is cyclosporine, and as in humans, nephrotoxicity can be an issue, particularly in rabbits where both acute and chronic nephrotoxicity have been observed as well as a distinct toxic syndrome following prolonged dosing [53]. Thus, while useful for examining short-term safety and efficacy, longer duration studies in immune-suppressed animals may be limited by significant systemic side effects, and use of control animals will be important.

Humanized mouse models may allow specific aspects of cell therapy efficacy and safety to be explored. A humanized mouse is generated from an immunocompromised mouse that is engrafted with human cells [54]. The reconstitution of the human hematopoietic system is one of the most advanced areas in humanized mouse research. The use of immunocompromised mouse strains such as NOD/SCID/gamma has greatly improved knowledge of human hematopoiesis, facilitating the development of multiple human cell lineages, including B and T lymphocytes, natural killer cells, dendritic cells, macrophages, and erythroblasts in the mouse. The models do have limitations, such as the rare differentiation of certain cell lineages from HSCs, immature differentiation, and insufficient intercellular relationships, but these models may be useful for examining the safety and efficacy of novel cellular therapies.

The selection of an appropriate animal model is challenging, and more than one animal model may be required to adequately address all nonclinical aspects of a cellular therapy program. The animal model choice should not be based solely on availability, familiarity, or cost but should be scientifically justified as providing the genetic, physiological, or pathological characteristics required to meet the study goals.

Table 5 **Characteristics of Immunodeficient Mouse Models**

Strain	Deficiency	Characteristics
Nude	• Thymus-derived T cells production is blocked (mice are athymic)	• The immune system is characterized by a small population of T cells, an antibody response confined to IgM class, a low T-dependent response to antigens, and an increased natural killer cell response. • Life-span 12 months • Predominant spontaneous lesions include hepatitis with associated systemic phlebitis, interstitial pneumonia with accompanying proliferation of bronchoalveolar epithelium, myocarditis, renal glomerular disease, and the development of spontaneous lymphomas
SCID	• Severe combined immunodeficiency affecting T and B cell development	• NK may be fully functional or partially deficient depending on strain • Varying reports on level of lethal thymic lymphomas • Life-span approximately 8–9 months • Predominant spontaneous lesions include thymic lymphomas and a lower incidence of nonthymic tumors including myoepitheliomas, rhabdomyosarcoma, and mammary adenocarcinomas
NOD/SCID	• NK cell dysfunction, low cytokine production, and T and B cell disfunction	• 10% of mice display immunoglobulin leakiness • Life-span approximately 6 months • Spontaneous lesions include lymphomas typically around 20–40 weeks with a frequency of >60% reported at 40 weeks and epicardial mineralization
NOD/SCID/ IL2Rγnull mice	• Lack mature T cells, B cells, or functional NK cells and are deficient in cytokine signaling	• Normal life-span • Low incidence of spontaneous lymphomas but typically with an earlier onset at 12–26 weeks with a higher incidence in female animals

6.3 The Three R's (Reduce, Refine, and Replace)

While it is acknowledged that animal studies may be the only method to obtain data to maximize the predictive value of the cell therapy nonclinical program for clinical safety and therapeutic activity, the opportunity exists for reducing, refining, and replacing (three *R*'s) animal use. The case-by-case approach to the design of nonclinical programs for the assessment of efficacy and safety of cell-based therapies is well placed to apply the principles of the three R's. Where there are no meaningful animal models, opportunities exist for the replacement of studies with well-designed in vitro alternatives. Where in vivo studies are relevant, the majority of work is performed in animal models of disease; it is therefore possible, using prospectively designed, protocol controlled studies, to obtain both efficacy and safety data from single studies, reducing the number of animals required (Section 6.2). The application of new imaging modalities such as whole body imaging potentially allows biodistribution study designs (Section 7.1) to be refined, removing the need for interim sacrifice groups and allowing each animal to be its own control.

7. Specific Nonclinical Safety Considerations

The overall nonclinical safety assessment of a cell-based therapy should be comprehensive enough to identify, characterize, and quantify the safety risks, including local and systemic toxicities and the reversibility of such toxicities. Traditional toxicology programs are typically of minimal value, and in many instances, the testing methodologies for a cell therapy are product specific. Detailed below is our current understanding on specific risk factors and issues to consider when designing studies to investigate these.

7.1 Biodistribution and Persistence

An important efficacy and safety consideration for cellular therapies is the biodistribution risk; where does the cell therapy go after it is administered? There is a need to understand the potential for trafficking, homing, and persistence of cells in both target and nontarget tissues. The potential impacts of administered cells distributing to a nontarget tissue include the possibility of off-target toxicities and the risk of cell engraftment at an abnormal tissue location (ectopic engraftment) although the risk of ectopic engraftment and its effects remain unpredictable. The distribution potential of a cell therapy will be impacted by the route of administration, the use of scaffolds and matrices, and whether the cell therapy exerts biological function via trophic mechanisms [55]. The identification of a small number of cells at an ectopic location or persisting longer than expected, however, does not in itself mean a halt to product development, but it identifies an issue that will warrant further investigation. Understanding the distribution and persistence of a cell therapy is a major technical challenge.

Following IV administration, cell therapies are often considered to distribute broadly throughout the body, but for some cell therapies under certain clinical conditions, this may not be the case. For example, some MSC therapies delivered systemically or exogenously in animal models of disease have been shown to home specifically to inflamed, damaged, and diseased tissues [56]. However, the number of cells homing to the disease tissue may be very low (less than 10% has been reported [57]), and a significant proportion of the systemically administered MSCs rapidly accumulate as emboli in the lung in animal models. Cultured MSCs can become up to 20 μm in diameter, which is considerably larger than the murine pulmonary microcapillary network, and it is, therefore, unsurprising that the MSCs become arrested. After 24 h, the number of cultured MSCs remaining in the lungs rapidly decreases, most likely through lack of cell survival, although some redistribution to other tissues such as the liver and spleen may occur [57]. Although administered cells will also pass through the lungs in the clinical situation, the extent to which they are retained may be significantly different, and clinical safety issues relating to lung emboli have not been commonly reported. A question, therefore, arises as to whether this biodistribution finding in mice adequately reflects what would be observed in the clinic and needs to be considered when assessing biodistribution particularly by the IV in small animal models.

Where extensive knowledge of a given class of therapy is available, the assessment of biodistribution and persistence potential may take the form of a scientific evaluation of the literature. However, specific cell modifications, such as genetic modification, or the cell culture conditions can alter the biological characteristics of a cell therapy and potentially impact the biodistribution potential, and this may limit the ability to use only a literature-based approach to assess the biodistribution and persistence risk. Cell culture duration, degree of expansion, and differentiation status may all alter the cellular characteristics and influence the biodistribution potential even for products that may appear similar. It is, therefore, important that supportive data from the literature and biodistribution studies have been or are performed with material that represents the intended clinical product.

It is essential for any in vivo studies that the cells are administered to the intended clinical location by the exact route that will be employed in the clinic including any devices or matrices. The persistence profile of a cell-based therapy may be very different if the therapy is part of an integrated device. A study in heart transplant patients, for example, demonstrated that MSCs in the transplanted hearts were all of donor origin even many years after transplantation [58], suggesting that under given circumstances, transferred MSCs can survive for extended periods of time, compared to the short-lived allogeneic MSC therapies administered IV. It is, therefore, possible that administering cell therapies directly into tissue or as part of integrated devices may alter their survival characteristics through the provision of a microenvironment that supports cell survival.

At present, there is no single satisfactory method of tracking the fate of administered cells in vivo. Limitations of biodistribution assays arise in terms of the assay sensitivity, the limits of detection, and in some instances in the animal models that can be used. The first challenge for the biodistribution studies is to ensure that appropriate conditions are provided to allow for cell survival. Human cells are xenogeneic in an animal host, and the potential exists for the induction of an immune response and subsequent cell loss. This may not reflect the clinical situation where cells may persist long-term, and therefore, the value of the biodistribution analysis is in question. Typically, the studies need to be run in an immunocompromised or immune-suppressed animal model; although, where data exists that a xenogeneic immune response is not induced, the use of normal (immunocompetent) animal models remains an option. The most complete immunocompromised animal models available are mice (see Table 5), but as discussed above, the cell therapy distribution pattern may not truly reflect the clinical situation as the larger human cells may become arrested in microvasculature.

Larger animal models can be used, but immune suppression may be incomplete, and the impact of this on cell survival must be considered as part of the data analysis. Although an autologous species-specific product could be used, the generation of such a product is not trivial, and differences between the cells are likely to exist [52]; therefore, if distribution was seen with the animal homolog, its relevance to the clinical product would not be clear.

Ideally, biodistribution studies will be run in a relevant animal model of disease, as the disease state may alter the risks of distribution and persistence. However, when the cell therapy forms part of an integrated product, this can be technically challenging, if not impossible, if the animal disease model is a rodent. In some instances, it may be possible to miniaturize devices, and as long as comparability can be shown, this route maybe scientifically justifiable. The use of large animal models, however, may be more relevant in certain circumstances, although this brings with it challenges with respect to immune suppression, biodistribution assay sensitivity, and potentially tissue handling, as will be discussed below.

Numerous methodologies have been employed to address the questions of cell therapy fate over time. Methodologies employed include single-photon emission computed tomography (SPECT) scanning, magnetic resonance imaging (MRI), bioluminescent imaging, and quantitative polymerase chain reaction (PCR). Some of imaging modalities enable cell detection within the living animal effectively, allowing the animal to be its own control, while other technologies will require animal sacrifices at multiple time points, tissue harvesting, and processing. However, the choice of methodology will be study specific and can involve a combination of both live animal imaging and post-sacrifice assessment. The utility of a given modality is determined by the sensitivity and limits of detection of the system.

Technologies that allow whole body imaging are becoming more readily available within the university and contract research facility setting. MRI is a noninvasive imaging technique that uses strong magnetic fields and radio waves to produce detailed images of the inside of the body and provides three-dimensional images with high resolution but is limited by low sensitivity. Positron emission tomography (PET) and SPECT are the main nuclear imaging techniques and can detect very low levels of radiolabel, making them attractive for cell therapy biodistribution studies in the nonclinical setting, but they have more limited spatial resolution. Typically, PET and SPECT are employed in conjunction with computed tomography (CT) and also more recently with MRI to provide more anatomical detail. To visualize the cell therapy using whole body scanning techniques requires the cells to be appropriately labeled: either direct or indirect labeling.

Direct labeling requires the introduction of an imaging agent into the cell therapy. For PET and SPECT, the cells are typically incubated with a radioactive label such as Indium (^{111}In) oxine [59]. It is important to determine the sensitivity of a given cell therapy to radiotracer labeling. Endothelial progenitor cells, for example, appear unaffected in terms of function, viability, and migratory properties following ^{111}In oxine labeling, whereas hematopoietic progenitor cells have been reported to be severely impaired [60]. For MRI, a contrast agent such as supermagnetic iron oxide particles (SPIOs) need to be delivered into the cell therapy product [59]. One disadvantage of

SPIOs is the signal from these can only be detected for a relatively short time period after cell administration, and SPIOs are, therefore, not useful for longer term studies.

Many cells, however, have poor uptake efficiency of contrast agents, and a variety of methods have been employed to improve this. Care has to be taken as to the method employed, however, to ensure that the cell functionality is not affected as some reagents have been reported to be toxic to the cells [61]. In addition, with both direct labeling options, it is not possible to distinguish live from dead cells. Phagocytic cells such as macrophages through engulfment of labeled dead cells may confound interpretation of biodistribution study data as these cells are known for their migratory properties.

Indirect cell labeling involves the genetic modification of the cell therapy product. Cells are typically transduced with a viral vector encoding a reporter gene. These reporter genes can mediate the uptake of radiolabels via membrane receptors or by acting as transporter proteins [61]. As with indirect labeling, the impact of the genetic modification on the cell function and viability must be assessed. In addition, any impacts relating to the cellular transformation need to be considered (see Section 7.2) and the potential for modification to increase the susceptibility of cells to an immune response. Unlike direct labeling, expression of the reporter gene is limited to living cells; in addition, the reporter genes are also incorporated into progeny daughter cells. Cells can therefore be, in principle, observed over their lifetime, and there is the potential for long-term persistence to be assessed.

A major limitation of whole body imaging is the minimum number of cells that can be detected. Reports suggest that the limit of detection may be as high as 10,000 cells [62]. This may not be sufficient to identify potential safety concerns resulting from the abnormal localization and persistence of a cell therapy. In addition, where biodistribution studies need to be performed in large animal models, the utility of whole body imaging is limited as signal attenuation due to tissue thickness is problematical.

Another option is intravital microscopy (IVM), which is a technique used to observe specific biological systems in vivo at high resolution. IVM requires access to the tissue of interest and a method for visualizing the cells, such as genetically engineering cells to express a fluorescent label such as green fluorescent protein (GFP). Before starting the experiments, the procedures have to be established to provide the conditions for the animals to tolerate the surgery and avoiding excessive physiological disturbances that could interfere with the results. Continuous improvements in spatial and temporal resolution in microscopy technologies, including strategies to minimize the motion artifacts caused by the heartbeat and respiration, provide the opportunity for imaging biological processes in multicellular organisms including mice [63]. This technology is of particular value for therapies where the route of administration is direct delivery to a specific organ such as the brain. Not only can the distribution and persistence of the cells be assessed within the local environment, but this can be combined with an assessment of physiological function. As with all techniques that require the genetic modification of the cell therapy, it is important to confirm that the function and viability of the cells are not compromised nor that the modification makes the cells more immunogenic, potentially affecting the distribution profile of the product and confounding data interpretation.

The alternative techniques for assessing cell distribution include immunohistochemistry (IHC) and qPCR. The major disadvantage of these techniques is the requirement for biopsy or animal sacrifice for the assessment of cell tracking. It is possible to combine these assessments with other studies; this may include terminal endpoints as part of whole body imaging study or endpoints in efficacy, pivotal safety, or tumorigenicity, thereby minimizing the use of animals and applying the concepts of the three R's (see Section 6.3). Where IHC or qPCR alone is the modality for assessing biodistribution, multiple sacrifice groups may need to be considered as part of the study design—early sacrifice groups to examine the immediate cell therapy distribution patterns and multiple later sacrifice groups to understand the long-term biodistribution and persistence of the product (Table 6). The biggest challenge for both qPCR and IHC is how to perform the tissue sampling. The tissues to be collected will be dictated by the route of administration. There is no fixed panel, and choice of tissues should be scientifically justified (Table 6). For large animal models and larger organs in smaller animal models, a tissue sampling strategy will need to be employed; a single sample from a porcine liver, for example, is unlikely to truly reflect the risk of a cell therapy distribution potential.

Assessment based on qPCR requires the identification and validation of DNA sequences specific for the human cell(s) of interest with no cross-reactivity to sequences within the test species cells. Sequences can include any imaging reporter genes that have been stably integrated into the genome (e.g., GFP) [64]. Alternatively, if the cell therapy source is of male origin and the test species are female, cells can be tracked using Y chromosome sequences [65]. Sequences selected for qPCR to identify the cell therapy must undergo documented validation to confirm they are cell specific. This will need to be confirmed for every tissue type that will be explored in the biodistribution study. The assessment of a multiple marker strategy may have utility for minimizing the risk of false positives, with all markers required to be positive for a true positive signal. Given that the cells will be of human origin, strategies will need to be developed and good sample handling practice employed to minimize the risk of cross-contamination of samples and the generation of false positive results. If the cell therapy under investigation is of male origin, for example, it may be possible to limit all tissue handling to female staff. The order in which tissues are collected can also minimize contamination. All control animals should have their tissue collected first. If the cells were delivered intramuscularly, the site of administration should be the last sample collected. Multiple changes of gloves are required per animal at tissue collection and comprehensive clean down procedures must be performed between each animal to minimize false positive data.

Immunohistochemistry is another imaging option, but it is essential to validate human-specific markers that can be used to identify single cells or small clusters of cells in test species tissues (e.g., Alu DNA sequences [66]; or human mitochondrial sequences [15]). To ensure that the test accurately identifies the cell therapy in all tissue types that will be assessed as part of the study, the IHC method must be optimized. Parameters to be considered include the expression levels of the marker of interest in the cell therapy and test species tissues and tissue fixing and processing methods. IHC assay optimization is typically performed by spiking tissue samples with the cell

Table 6 Example Study Designs

Study	Study Parameters[a,b]	Considerations
Determine cell fate/ biodistribution and persistence Study does not need to be run to GLP but feasibility should be considered if combined with the tumorigenicity assessment	Species/strain	• Large versus small animal model • Immune-competent versus immunocompromised or immunosuppressed model • Ability to deliver by the intended clinical route of administration using the clinical delivery method including scaffolds • Disease is likely to impact the distribution and persistence and animal models of disease should be used where appropriate • Impact of immune system on distribution profile may not be feasible • Imaging modality to be used
	Duration of in-life phase	Dependent on predicted persistence of products: • For transient products (e.g., IV-delivered immunomodulatory MSC therapies) duration should confirm clearance typically 4–12 weeks duration • For multi-dose products or products expected to engraft 3–12 months (dependent on product and the life-span of animal model, persistence of less than 1 month may only be feasible for acute disease models) • Studies may also provide information on tumorigenicity risk and combined tumorigenicity/ biodistribution studies can be run
	Group size	• Murine animal model (minimum of 5–10/group) • Large animal model (3–5/group)
	Dose	• Clinical dose or maximal feasible dose, justification for dose required
	Test groups	Use of in vivo imaging technologies may allow the number of interim sacrifice groups to be limited. Time of interim sacrifice time points will be dictated by the expected persistence of the product • Untreated control • Clinical product—sacrifice time point 1 (e.g., 1 week post dose) • Clinical product—sacrifice time point 2 (e.g., 2 weeks or 3 months post dose) • Clinical product—sacrifice time point 3 (e.g., 4 weeks or 6 months post dose)
	Measurements	• In vivo imaging if available/relevant • Body weight and clinical observations over duration of study
Combined safety/ tumorigenicity assessment	Species/strain	• Tissue list at study termination, product specific but can include: site of administration, brain, bone marrow, kidney, liver, lung, lymph nodes, spleen, heart, gonads, gross lesions, other therapy specific target tissues • Validated cell detection assay e.g., qPCR and/or IHC

...studies to ...GLP should be considered.	...s processes
phase	• 6–12 months for studies where there is a risk of pluripotent cell contamination • 3-month studies may be acceptable for other cell therapies with supporting in vitro data
Group sizes	• Pilot studies may provide information on sex differences to risk of tumor formation and sensitivity of model helping to determine group sizes • A minimum of 10/group is recommended and for longer duration up larger numbers of animals may be scientifically justified to allow detection of rare events
Test groups	• Cell therapy vehicle (negative control) • Clinical dose or equivalent group • Interim dose groups can be considered • Positive control group(s)
Measurements	• Perform standard toxicology assessments (mortality, behavioral observations, body weights, feed consumption, ophthalmology) and clinical laboratory parameters • Full necropsy, for studies incorporating toxicology the standard full tissue list should be considered (see Bregman et al., 2003[c] for list), this can be adapted based on scientific justification on a per-study basis • Macroscopic and microscopic evaluation of tissues • Selected tissues including site of administration and all proliferative lesions will be assessed by validated assay to determine presence of cells from human origin • The species of origin of neoplasms (murine or human) should be assessed
Immunogenicity assessment	In vitro assessment[d] • Expression of immunological surface molecules on cell therapy product (+/− presence of inflammatory cytokines) by flow cytometry • Analysis of secreted chemokines and cytokines (+/− presence of inflammatory cytokines) using ELISA-based methods • T cell proliferation assay • Susceptibility of product to natural killer cell-mediated cytotoxicity • Risk of human serum antibody-mediated lysis In vivo assessment • Immune-competent animal strain • Can be combined with efficacy and/or biodistribution studies • Serum samples for assessment of anti-cell therapy antibody response and cytokine analysis • Splenocytes or PBMC for assessment of cell-mediated toxicity

[a]There are no standard study designs, case-by-case assessment is critical and negotiation of study designs with regulatory authorities is essential.
[b]Adapted from Sharpe et al. Nonclinical safety strategies for stem cell therapies. Toxicol Appl Pharmacol 2012; 262(3):223–231.
[c]Bregman et al. Recommended tissue list for histopathologic examination in repeat-dose toxicity and carcinogenicity studies: a proposal of the Society of Toxicologic Pathology (STP). Toxicol Pathol 2003; 31(2):252–253.
[d]Okamura et al. Immunological properties of human embryonic stem cell-derived oligodendrocyte progenitor cells. J Immunol 2007; 192:134–144.

therapy material and confirming the ability to identify the cells. Untreated tissue will provide information on the level of nonspecific false positive results, and this may differ with different tissue type.

Currently there is no one single satisfactory method for assessing biodistribution, and a pragmatic approach is recommended. Multiple imaging modalities including both whole animal imaging and tissue-specific sampling may be required to understand risks associated for a given product.

7.2   Genetic Modification

Genetic modification is the process of modifying or inserting a new genetic sequence into a cell [67]. Although a number of different technologies can be employed including nonviral methods and the use of adenoviral vectors, retroviral and lentiviral gene transfer systems are the most commonly employed in the genetic modification of cell therapies since vectors capable of sustained high levels of expression and the ability to package large inserts are required. Vector systems derived from these *Retroviridae* family of viruses come with two generally accepted risks: the production of replication-competent viruses (RCV) and insertional mutagenesis, specifically oncogenic activation. However, these risks are mitigated by the use of replication-incompetent viruses and self-inactivating vectors (SIN).

The issue of reducing the probability of RCV has been systematically addressed during the development of retroviral vectors, with each new generation aimed at minimizing and reducing the risk. Modern retroviral vectors have the most advanced split-packaging design. The viral packaging elements are on separate plasmids to those encoding the sequences for reverse transcription and viral integration. These split-gene packaging strategies [68] reduce the risk of generating RCV because multiple recombination events are necessary to create a virus that harbors the sequences required for independent replication. Additional safety measures have been introduced through the generation of SIN vectors [69]. SIN vectors contain deletion in the enhancer region of the 3' U3 region of the long terminal repeat, which results in a transcriptionally inactive vector that cannot be converted into a full-length RNA, hence reducing the risk of replication-competent virus formation. Risk is, therefore, linked to the generation of the retroviral vector used, with earlier first- and second-generation vector systems having the greater risk of recombination events that could lead to RCV.

The results of replication-competent retrovirus testing from clinical trials of genetically modified T cell therapies carried out at the Baylor College of Medicine, the St. Jude Children's Research Hospital, the National Cancer Institute, the University of Pennsylvania, and Indiana University have recently been reported [70]. Thirty master cell banks (MCB) were generated and 42 viral lots were produced. RCV screening of the MCB and viral supernatant lots were consistently negative. The number of genetically modified T cell therapy products generated was 297, and of these, at the time of publication, 282 products were negative for RCV with 15 product results pending. Testing was performed on 629 clinical samples ranging from one month to eight years post-infusion, and so far, RCVs have not been detected in a single patient, suggesting that the risk is low.

The probability of formation of RCV and potential recombination with retroviral sequences of the target cells with the later generation of retroviral vectors is still theoretically possible, but it is currently assumed that careful engineering of these systems has led to the point where they are free from RCV, which should no longer be a major safety issue. However, regulatory guidances recommend an assessment of the risk of the presence of replication-competent viruses in the product. This is typically performed as part of the quality assessment of the product.

The insertion of retroviral vectors into the host genome has meant that insertional mutagenesis is a risk for genetically modified cell therapies. Insertional mutagenesis can occur through activation and silencing of genes or dysregulation. Retroviral insertion into host cell DNA occurs with high efficiency but without clear preference for specific DNA target sequences or loci; therefore, each transduced cell is clonally marked by the specific insertion site within that cell. The preferred target of the gamma retroviral integration machinery is open chromatin, which is associated with transcriptionally active regions and, therefore, is dependent on cell cycle and differentiation [71]. Even lentiviruses, which are capable of active nuclear transport into cells resting in the G1 phase of the cell cycle, prefer active genes for insertion but without an obvious preference for promoter regions [72]. Insertional analysis only reflects endpoint scenarios and those that persist within the cell therapy to the time of analysis. Causal mechanisms underlying preferential integration are still not resolved. For any type of vector, the genotoxic risk increases with both the number of cells manipulated and infused into the recipient, the vector copies per cell, and the number of insertion events per cell.

Oncogenic activation has been observed in the clinic following the administration of gamma-retrovirally modified HSC, with leukemias or pre-leukemias reported in three clinical trials [73]. In general, the risk of insertional mutagenesis, while poorly defined, is considered to be related to disease background, cell type to be transduced, and vector characteristics [74]. Numerous clinical trials with gamma-retrovirally modified T cells have not yielded evidence for insertional adverse events despite long-term persistence of transduced cells [74], and lentiviral vectors have not yet been associated with insertional oncogenesis, although integration-mediated clonal dominance has been reported in one trial [75]. These data suggest that disease background factors and cell-intrinsic mechanisms may modify the risk of insertional mutagenesis.

An assessment of the risk of insertional mutagenesis should be completed for a genetically modified cell therapy. A number of new in vitro assays and animal models are available that can be used to predict the risk of insertional mutagenesis [76]. In vitro assays using immortalized C57BL/6J bone marrow cells can be used to determine the incidence of mutants based on the number of cells that need to be exposed before development of a transformant. Standard processes, however, must be developed as culture conditions and cell densities can impact reproducibility of results. In vivo studies can be performed typically in the mouse (strains used are in vivo model specific). Study endpoints include preferential survival of a specific subset of genetically modified cells (clonal skewing), tumorigenesis/leukemogenesis, the relative abundance of a genetically modified cell (clonal fitness), or integration site analysis depending on the model chosen. But each model has limitations, not the least the low sensitivity and inability to deliver a clinical dose to the murine animal models. These factors should be considered before committing to a specific study, and interaction with the regulatory agencies on your proposed study plans is highly recommended [77].

7.3 Tumorigenicity

An important safety consideration for any cell-based therapy is the risk of tumor formation. Although predominantly considered in respect to pluripotent cell-derived therapies, all cell-based therapies including somatic cell therapies have the theoretical potential to form tumors. Tumorigenic potential can be influenced by many factors including the type of cell therapy, the differentiation status, and proliferation capacity of the cells, whether the cells are genetically modified (see Section 7.2), the phenotypic plasticity of the cells, the intended clinical location, and the long-term survival of the product.

Since genetic aberrations have been strongly associated with cancers, it is important that cell therapy preparations destined for clinical use are free from cancer-associated genomic alterations, and this requires defined culture conditions and the cytogenetic and genetic characterization of the final clinical product. The continuous culture of a cell therapy can provide the selective pressure for genetic change. For pluripotent cells, as well as aneuploidy, a number of other subchromosomal changes have been identified [32,33]. It is possible that many of the chromosomal changes observed in the pluripotent cell cultures are adaptive and confer a proliferative advantage to the cells. However, some of the aberrations identified in iPSC lines are suspected to originate from the parental somatic cells [32], or the aneuploidy was evident in early passage of the iPSC, suggesting considerable selective pressure during the reprogramming process, irrespective of the type of reprogramming used [78]. Genetic stability issues have also been reported for MSC lines; DNA losses and gains, DNA methylation instability, and evidence for telomeric deletions in subpopulations of cells have all been observed during culture to late passage [79].

For pluripotent-derived cell therapies, one aspect of tumorigenic risk is the potential for the presence of small numbers of undifferentiated cells within the final product. The ability of the human immune system to identify pluripotent cells as immunological targets is unknown. Evidence exists that there is T cell reactivity to pluripotency markers such as Oct4 in healthy donors [80], indicating that in a patient with a healthy immune system, the risk of tumorigenicity from rare contaminating pluripotent cells may be mitigated. However, this cannot be assumed for all patients, and the risk of tumorigenicity, in particular teratoma formation, needs to be addressed.

Methods are in development to eliminate the undifferentiated cells from the final pluripotent-derived cell therapy product; including the use of pluripotent apoptotic agents, stage-specific genotoxic agents, activated cell sorting, and the use of monoclonal antibodies against undifferentiated stem cell surface markers [81,82]. No method has been approved, and heterogeneity of the clinical product cannot yet be excluded. And, engraftment of undifferentiated or incorrectly differentiated cells may still present a substantial tumorigenic risk to the recipient, and therefore, tumorigenicity assessment will still be required.

The risk of tumorigenicity may be assessed by in vitro and/or in vivo assays; in addition, published data on related products may provide additional supportive information, although the relatedness of the products will need to be determined. A number of different parameters may define a cell as having tumorigenic potential, and some of these may be assessed using in vitro assays. Growth rate, propensity for differentiation, and population doubling times should be monitored during product development, and any changes investigated further, in particular monitoring for clonogenicity. An assessment of telomere length and chromosomal abnormalities should also be monitored during product development, particularly for cells that are maintained over multiple passages. Screening cells via molecular biology methods such as qPCR for specific somatic alterations in cancer-related pathways can be considered, particularly those related to tumors associated with the final differentiated product. These can include mutations in tumor suppressor genes or genes upregulated during epithelial to mesenchymal transition that have been associated with invasion and metastasis of malignant cells. Transformed cells also have a unique characteristic, which is the ability to multiply without adhering to a solid support [83]. This forms the basis of the soft agar assay that can be used to determine anchorage-independent growth potential. Appropriate controls should be used in all assessments and the limits of detection of the assay understood.

For some products, a paper-based scientific rationale and/or in vitro studies may provide sufficient information to assess the tumorigenic risk of a cell therapy product. For other products, an in vivo assessment will need to be performed. The nonclinical in vivo studies to investigate the risk of tumor potential of a cell therapy must be designed to allow assessment of rare events. These studies present a number of challenges including the selection of the optimal animal model, a balance of a feasible group size with statistical power, study duration, dose, and route of administration. In addition, the requirement for special husbandry and care for immunodeficient animals to minimize loss of animals to opportunistic infection exists, particularly for studies of up to one year in duration. Consideration needs to be given to how, or if, these studies can be performed in a GLP-compliant manner.

Tumorigenicity studies must be performed with the intended clinical product. Animal analogous products are rarely acceptable for the assessment of tumorigenic potential as species-specific differences may alter tumorigenic potential. The cell therapy product should be manufactured using the intended clinical manufacturing protocol, as cells produced under differing conditions may have different molecular and growth characteristics altering the tumorigenic risk.

Since the product is typically of human origin, the studies require the use of immunodeficient or immune-compromised mice. Studies have shown that tumors develop more consistently and rapidly in mice deficient in T-lymphocyte, B-lymphocyte, and NK cell function than mice that lack function of only one or two of these cell types [84]. The choice of immunodeficient mouse will be specific for a given program and will be driven in part by the characteristics of the mouse strains, including life-span and spontaneous tumor burden as a function of age and degree of tolerance for human cells (Table 5). Although it is possible to use chemically induced immunosuppression for chronic in vivo assessment of tumor formation, immunodeficiency is inconsistent, and the study animals may be compromised by concurrent drug treatment [84]. In addition, the impact of the immunosuppressive drugs on the functioning of the cell therapy must be understood, particularly when the cell therapy product has immuno-modulatory action.

Small-scale pilot studies may provide valuable information and can help inform the design of the final pivotal tumorigenicity study. Some groups, particularly those working with pluripotent cell-derived products, perform a pilot study to assess the sensitivity of different animal models to their particular starting pluripotent material. The pivotal tumorigenicity studies are looking for the risk of contaminating starting material in the final product, and the pilot dilution series studies are performed to identify the minimum number of potential contaminating cells that can be detected to induce tumor formation [84]. These studies can also be used to explore methods to optimize study output. Injecting small numbers of undifferentiated pluripotent cells with irradiated human fibroblast feeder cells has been shown to increase the sensitivity of in vivo tera-toma assays. It has been reported that as few as two cells can induce teratoma formation under these conditions compared to 1×10^4 cells cultured under feeder-free conditions [84]. These differences are likely real and may reflect a role for fibroblasts in reducing cell death at the time of injection. The pilot studies will also allow technical difficulties to be identified and confirm cell engraftment at the administration site.

The cell therapy should be, where possible, administered by the intended clinical route of administration to the intended site of delivery. Local environment at the delivery site may significantly impact tumor potential—from the number of tumors that form to the rapidity of tumor growth. However, given that these studies are typically run in immunodeficient mice, the impact of the immune response cannot be assessed. Data suggest that the immune system, in particular natural killer cells and the complement system, may be able to reject small numbers of pluripotent cells, mitigating the tumorigenic risk [85].

The site of administration particularly in a mouse can impact the cell dose that can be administered. Although it is desirable to assess the clinical dose and potentially several dose levels via the intended clinical route, this may not be possible. Options include administering a maximum feasible dose and providing within regulatory submissions the rationale for its relevance to the clinical setting. Alternatively, an additional group or groups may be included that use an alternative route of administration such as subcutaneous or intramuscular delivery; this may allow the delivery of the full clinical dose. Some have argued that these routes may be more sensitive and reproducible [84] but may not assess the impact of an appropriate local environment. Caution must also be taken when considering a combined biodistribution/tumorigenicity study as the route of administration can alter the biodistribution profile of a product.

A key consideration for the study design is the choice of control groups (Table 6). Vehicle controls should always be included and provide information on background incidence of tumor formation in the test species. Positive control groups are also recommended. For pluripotent cell therapies, this should be the original undifferentiated parental stem cell line. But for other cell therapies, the choice is not so simple. In principle, it should be an appropriate tumor cell line. Aggressive tumor cell lines should be avoided as they may not appropriately assess the susceptibility of the animal model to a less tumorigenic cell line. For some therapies, partially differentiated cells may also be an appropriate control.

The number of animals to use in each treatment group should be adequate to ensure statistical significance of any biological observations. But, what does this mean in practice? Reported studies have utilized between 5 and 20 animals per group. A balance, however, needs to be achieved between a practical study size and the need to investigate the potential for the likely rare event of a tumor formation. A fully transformed cell line will actually give a clear positive result with very small numbers of animals. Availability of clinical-grade product can also impact the practical group sizes that can be achieved. A justification for the planned group size should form part of the discussions with the regulatory agency on the proposed nonclinical package for a given product.

The duration of studies exploring tumor formation should be sufficient to allow identification of potentially rare events (Table 6). Reported durations of studies have been as short as 3 months but typically 6–12 months in mice and up to 20 months in a rat model [43]. Duration may be dictated by the life-span of the model. Many lines of immune-deficient rodents have survival rates of less than 1 year, and many models over 6 months of age develop spontaneous neoplasms that could confound interpretation of study data [15]. Studies of a minimum of six months are typically run for pluripotent-derived products. Three-month studies may suffice for adult cell therapies when scientifically justified, due to confirmed lack of persistence of cell product, but a minimum of 6-month study durations may be required where cell engraftment of the product is a possibility.

During the in-life phase of the tumorigenicity study, animals should be observed for specific endpoints including clinical observations, body weights, and where applicable, palpation for mass formation. Clearly defined humane endpoints for euthanasia will be required particularly if administration is to volume-limited organs such as the eye. Where early sacrifice is likely to be required, consideration should be given to when the supporting untreated control animals should be sacrificed to ensure age-matched controls are available for histological assessment. Complete necropsy should be performed with histopathology and other special analysis of selected tissues. Histology should not be limited to the site of administration but include a representative tissue list (Table 6) so that both tumorigenic potential and possible toxicity may be assessed. The tissue list may be informed by any prior biodistribution studies; tissues that have been found to contain cell product should always be included in the final histology tissue list.

Spontaneous proliferative lesions are common in immune-deficient animals and may confound study interpretation. This highlights the importance of negative and/or vehicle control groups. In addition, there may be published data on the background incidence of tumor formation for a given strain. Together, this information will provide study context. Spontaneous thymic lymphoma that progresses to systemic distribution is a common pathological finding of the NOD/SCID mouse strain. This was observed during the safety assessment of placental-derived mesenchymal stromal cells [86]; these findings occurred in both test article and negative control groups, providing confidence that these were spontaneous lesions. If a lesion is found in a given tissue in one animal, it is highly recommended to examine that tissue for all animals, even if the lesion was in the control groups. It is recommended that all tumors and proliferative lesions should have their cellular origin identified (human or host) via IHC analysis using a human-specific marker (e.g., Alu DNA sequences [66]; or human mitochondrial sequences [15]) or PCR analysis (human-specific genes, e.g., human GAPDH) using validated assays that can detect human cells in a range of tissue types.

It is possible to combine tumorigenicity studies with the toxicology and/or biodistribution studies. This may require additional groups as multiple sacrifice time points are generally required in these latter studies, but in some cases, group sizes may be smaller. Considerable thought must be given to the final design to ensure that no one aspect is compromised. Guidance from regulatory authorities should always be sought prior to initiating complex chronic studies.

Given the variable nature of scientific knowledge and clinical experience, a risk-based approach can be applied. Pluripotent cell-derived cell therapy products have to date routinely included in vivo studies of tumorigenicity due to their inherent risk profile and the early stage of their clinical translation. However, recently published data by the FDA has shown that across all cell therapies, only in 43% of submissions were tumorigenicity assays performed by testing a product directly (in vitro/in vivo), and that in 57% of cases, tumorigenic potential was assumed based on "consideration of product attributes, literature and/or previous clinical experience" [46].

7.4   Immunogenicity

Risks arising from the induction or modulation of the immune response can be separated into two distinct classes: the induction of an immune response by the patient against the cell therapy with the concomitant risk of reduced efficacy of the product or the induction of an immune reaction by the product on the patient resulting in toxicity. Assessing the risk of immunogenicity and cell therapy-induced immune-mediated toxicity can be challenging nonclinically, and extensive knowledge on the likely mode of action, product characteristics, production methods, and nonclinical assessment limitations will be needed to provide the context for study findings.

7.4.1   Immunogenicity

Immunogenicity is likely to be a major barrier to the successful development of allogeneic cell therapies. Immunogenicity may be influenced by multiple factors, including the allelic differences between the product and the patient, the relative immune privilege of the site of administration, the maturation status of the cells, the need for repeat administration, the immunological basis of disease, and an aged immune system. Experience with HSC transplantation shows that the immune system is exquisitely tuned to identify cells as foreign [87]. Rejection occurs due to allelic differences at polymorphic loci of the different transplant antigens: the ABO blood group antigens, human leukocyte antigen (HLA)/major histocompatibility complex (MHC) antigens, and minor histocompatibility (mHC) antigens.

Immune privilege has been proposed for pluripotent cell-derived therapies, however, the majority of the reported data has examined undifferentiated and early differentiated products [88]. These products may be immunologically immature, and the study findings may not truly reflect the rate of rejection a fully differentiated and immunological mature product could encounter in the clinical situation. Expression of immunological markers such as MHC class I increase on product maturation potentially increases the risk of an allogeneic immune response [88]. Many of the published nonclinical in vivo studies have focused on mouse pluripotent cells transplanted into a murine host in an allogeneic situation. Whether the findings can be extrapolated to the clinical situation with a differentiated product seems doubtful. The impact of MHC class I upregulation upon cell maturation has been investigated comparing the survival of undifferentiated and differentiated mouse ESCs when transplanted into an allogeneic host. Differentiated cells were immunologically rejected within 14 days, compared with the 28-day immunological rejection of the undifferentiated product [89]. It has also been shown that there is a significant increase in MHC class I expression following the formation of the embryoid body, a common procedure in pluripotent cell differentiation suggesting that for pluripotent cell therapies, MHC class I expression is dynamically regulated [88].

Immunogenicity of allogeneic pluripotent cells has also been studied in a model of myocardial infarction. Administration of mouse ESCs into the myocardium of allogeneic immunocompetent mice has resulted in a robust inflammatory response, with cellular infiltration by both innate and adaptive components of the immune system [90]. The immune responses were progressive and correlated with increased expression of MHC class I antigens on the transplanted cells. If a cell therapy is antigenic, a memory immune response will be generated that could lead to a rapid and robust response on subsequent exposure, limiting the opportunity for repeat dosing. Further studies have indicated that the immune response generated may be sufficient to prevent the long-term engraftment of allogeneic cell therapies. It is, therefore, likely that the immunological characteristics of a hESC-derived cell therapy are in constant flux depending on the differentiation status of the product [91]. This emphasizes the importance of assessing the immunogenic potential of the final cell therapy product and where possible using in vitro or in vivo conditions that mimic the clinical environment into which the product will be delivered.

The most comprehensive assessments of immunological responses to a cell therapy have been performed on MSCs. MSCs exhibit marked immunomodulatory activity through multiple mechanisms including the reduced expression of surface markers, via direct cellular interactions and through soluble factors [20]. In vitro, MSCs have been able to silence each aspect of the cellular rejection process. The in vitro evidence supports that MSCs exhibit marked immunomodulatory activity through multiple mechanisms, although it does not necessarily follow that there is a direct translation to preventing allogeneic therapy rejection in an immune-competent patient. Allogeneic MSC administration into immune-competent animals has generated evidence of anti-allogeneic MSC antibody generation and T cell responses [22,92]. The level of immune response generated was study-specific—in some studies, the responses were weak, but in other studies, the MSCs were strongly immunogenic, sensitizing against subsequent repeat administration [22]. This may reflect differences in study methodology, but the full characteristics of the MSCs were not reported and may, therefore, also reflect functional differences in the MSC preparations. Although allogeneic MSCs have the capacity to initiate both cellular and humoral responses in vivo, these may be considerably attenuated compared with other allogeneic cell types potentially delaying the time of rejection and providing a window of therapeutic benefit but limiting the option for subsequent re-exposure.

Cell therapies are by their very nature species-specific. It is not possible to fully assess the risk of an allogeneic response to a clinical cell therapy product in the nonclinical setting. Although a species-specific product may allow some immunological aspects to be assessed, the species-specific product and the host immune response may have differences that limit the translatability. The induction of a xenogeneic immune response would indicate that a cell therapy has the potential to be immunogenic; however, it is possible that a xenogeneic response may be stronger than the subsequent allogeneic reaction in humans and would, therefore, not be confirmatory that such a response would occur in the clinical setting but would highlight the risk. Lack of xenogeneic rejection in an immunocompetent animal may provide a degree of confidence that the cells may be less immunogenic in the clinical situation.

Another factor that can influence the immunogenic potential of a therapy is the clinical environment. Many cell therapies are administered to an inflammatory environment, characterized by the upregulation of disease-specific pattern of cytokines. Many murine cytokines (e.g., interferon gamma, interferon alpha and interleukin-6) show species specificity [93], and this may modulate the impact of the nonclinical environment on the clinical product. In addition, in many studies, the animals will be immunocompromised or immune-suppressed to allow dosing of the human clinical product, limiting the information that can be obtained on the effect of the environment on the product immunological characteristics. One option is to consider a comprehensive panel of in vitro immunogenicity studies to examine the potential of the cell therapy to act as targets for both innate and adaptive immune effector cells [94] (Table 6). These studies examine cell surface markers, chemokine and cytokine expression, and susceptibility to cell-mediated and serum cytotoxicity. In addition, the in vitro assessments can also confirm that the cells maintain the inherent immunogenicity characteristics of the cell type. While not fully predicting the immunological consequences after transplantation, they provide surrogates of potential clinical risks.

For a stem cell therapy derived from an individual's own cells, the risks of immune rejection due to allelic differences are low. However, cell culture and environment and selective pressures, such as forced gene expression that occurs during the production of iPSC lines, could theoretically alter the immunogenicity profile of a cell therapy [95]. Culture conditions should be clearly defined and the potential impact on the immunogenicity profile of the cell therapy recognized. Normal human sera contain antibodies against the sialic acid-derivative Neu5Gc, which is present on most mammalian cells excluding man, and can mediate lysis of hESCs that were grown on an irradiated mouse embryonic fibroblast feeder layer [96]. Other culture media components have also been implicated in graft-induced immune responses including bovine serum albumin (BSA). In a clinical trial of an MSC therapy for children with osteogenesis imperfecta, one subject developed antifetal bovine serum antibodies [97]. The antibodies were most likely against components derived from the cell therapy culture media and resulted in a systemic febrile response in the subject on repeated dosing. Thus, the potential exists for unexpected immune responses minimizing a therapy's effectiveness that alterations in culture conditions could address.

Immunoisolation through the use of integrated products such as encapsulation devices have shown promise nonclinically in preventing graft rejection in the absence of immunosuppressive drugs. However, the impact in the clinical setting has been less clear. Encapsulation typically involves placing cells within a semipermeable inert membrane. These have a pore size that allows small molecules to pass through but prevents the passage of T cells and antibodies. However, it has been found that chemotaxis of small molecules from the cells can attract and activate macrophages, resulting in the release of proinflammatory cytokines that are sufficiently small, so they too can pass through the membrane and potentially compromise the cell therapy [98].

There have been concerns raised over the influence of the genetic background on the reprogramming process and the potential to alter the immunogenicity profile of a reprogrammed cell. Studies in mice have suggested that retroviral and episomally derived iPSCs from C57Bl/6 mice were rejected after transplantation back into C57Bl/6 mice [95]. The reprogramming efficiency is not clear nor were many clones tested, and variation among the iPSC clones due to partial reprogramming cannot be excluded. These studies, however, do indicate that the absence of immunogenicity cannot be assumed for a reprogrammed cell therapy and should be addressed as part of the nonclinical safety assessment.

The immune privilege of the site of administration of a cell therapy is often cited as providing protection for an allogeneic therapy from an immune response. Immune privilege does not mean the absence of an immune response but rather immunological control through the expression of local factors such as transforming growth factor-β (TGF-β) to prevent uncontrolled and potentially catastrophically-damaging immune responses in tissues such as the eye and brain. If the general integrity of the tissue remains, there is the possibility that reactions to an allogeneic therapy may be attenuated, providing a window for therapeutic benefit. In many instances, however, the diseased state may compromise the immune privilege through the loss of tissue–blood barriers, and the allogeneic cell therapy may be subject to the full force of the immune system. In addition, the process of administering cells can induce inflammatory reactions that may lead to loss of cells or loss of cell function, limiting the efficacy of the product.

7.4.2   Cell Therapy-Induced Immune-Mediated Toxicity

Cell therapy-induced immune-mediated toxicity is critical to consider for cell therapies that primarily function in or modulate the immune system such as MSC therapies and genetically modified T cell therapies. However, the potential for cell therapy-induced immune-mediated toxicity needs to be considered for all cell therapies and will be associated with the proposed and known functions of a given cell type.

Cells such as MSC have marked immunomodulatory activity, and this is the proposed therapeutic mode of action of many of the products in development. The targets for this immunomodulation include dendritic cells, regulatory T cells, natural killer cells, T helper cell differentiation, B cell/plasma cell activation, and antibody production. For a given disease indication, an assessment must be made to confirm that there is not the potential for an adverse impact of the immunomodulation. To date, none has been reported, and the safety profiles are good, but the risk still needs to be assessed.

Cell therapy-induced immune-mediated toxicity has been reported for genetically modified T cell therapies being developed for the treatment of cancer. There are three potential immune-mediated toxicity risks: on target off tumor activity, off tumor reactivity, and cytokine release syndromes [99]. T cells play a key role in cell-mediated immunity, and recently, strategies to genetically modify T cells either through altering the specificity of the T cell receptor (TCR) or through introducing antibody-like recognition in chimeric antigen receptors (CARs) have shown promise in clinic for the treatment of cancers [42]. The challenge in the development of a genetically modified T cell therapy is the selection of the antigen target.

Although some tumor antigens are novel, many are overexpressed antigens that are also found at low levels on normal cells. The genetically modified T cells, therefore, have the potential to trigger potent cellular immune responses against these other tissues as well as the tumor. This type of reaction is known as on target off tumor reactivity. Early promising clinical data has been reported for the treatment of B cell leukemias using T cells that have been engineered to target the CD19 antigen, which is present on the surface of nearly all B cells, both normal and cancerous [3,4,51]. The therapy has been very successful at eliminating cancerous cells but is also causing the continuous depletion of normal B cells. Unlike normal small molecule or biological therapies, the T cells persist for long periods of time with the potential for lifetime presence. Although in this case, the on-target off=tumor toxicity can be managed by immunoglobulin transfer, it highlights the challenges of identifying tumor-specific targets. When the T cell antigenic target is a peptide, as in the case of genetically modified TCR T cell therapies, the challenge can be even greater; as well as being present in the desired antigen target, the peptide may also be present in unrelated proteins. Another undesirable reaction that can occur is off-target reactivity, and this has also been reported in clinical trials. This cross-reactivity is particularly a risk for genetically modified TCR T cells, which may react against related peptides in proteins other than the targeted ones. The consequences of an immunological response against a nontarget tissue can be fatal [100].

Extensive screening studies, therefore, need to run to assess the risk of on-target off-tumor reactivity. These include determining the expression pattern of the antigen in other tissues. This may consist of information from the literature, nucleotide and protein sequence database searches, and tissue cross-reactivity studies. Confirming expression patterns in tissue in vitro can be challenging and may require the use of more biologically relevant cell culture systems (Box 6). For the antibody-like recognition of surface molecules by CAR T cell therapies, animal models may have utility in assessing the risk of off=target toxicity, where there are limited differences in protein sequences of the species-specific proteins and tissue expression patterns are similar. For genetically modified TCR T cell therapies in the majority of cases, animal models will be unsuitable for predicting off-target toxicity because of the differences in gene expression and peptide presentation between the species.

Box 6 Identification of a Genetically Modified TCR T Cell Therapy Cross-Reactive Target

Elegant studies by Cameron et al. [49] examining the off-target toxicity of a genetically modified TCR T cell therapy targeting a MAGE-A3 peptide identified a peptide from the protein titin as the potential cause of fatal on-target off-tumor reactivity in the heart identified in a clinical trial. Standard in vitro cell culture assessments failed to show the expression of titin in cardiac myocytes. Following co-culture of the genetically engineered T cells with a set of 38 normal cardiac-derived primary cells, no T cell activation activity (IFN-g ELISpot assay) was observed against any of the cardiac cells. Indeed, the ability of the genetically modified T cell to target cardiac cells was only able to be confirmed when the team used a more biologically relevant cell culture, the iCell cardiomyocytes, which contain a mixture of spontaneously electrically active atrial, nodal, and ventricular-like myocytes with typical cardiac biochemical, electrophysiological, and mechanical characteristics. Co-culture of the genetically modified T cells with the iCell cardiomyocytes resulted in cell killing.

Some safety considerations relate directly to the function of a cell therapy and, in particular, the effectiveness of that function. Recent clinical data have shown that the T cell therapies can be very effective against the target tumor by inducing tumor cell lysis and potentially tumor cell removal at a faster rate than is seen with traditional immune therapies. This can result in high levels of cytokine release and macrophage activation syndrome [3]. These types of reactions are very difficult to replicate and predict nonclinically, but the potential impact of uncontrolled or high efficiency of function should be considered as part of the overall risk assessment.

7.5 Toxicity

A general assessment of toxicity needs to be made for a cell therapy product. The overall design of the toxicology studies should include the basic tenets used in all toxicology studies, including clinical dosing route, dosing schedule, and where applicable, multiple doses (Table 6). Studies should also include adequate numbers of animals per group, although the numbers may vary depending on the safety concerns for a given cell therapy product, the animal species, the disease model, and the delivery system used. Use of animals of both sexes should be considered, as the different sexes can have differing susceptibility to toxicities. Where dosing for a whole study cannot be completed on the same day due to the complexity of the dosing procedure or timing of product administration to disease status of the disease model, then considerable thought must be given to how the study will be dosed to minimize study bias as much as possible. There should be appropriate use of control animals. This may include untreated animals or animals that receive a formulation vehicle only or scaffold alone; the rational for choice of control groups should be provided in study documentation.

Control groups allow the study findings to be appropriately interpreted particularly if historical control data for an animal disease model is limiting. For the pivotal safety assessment, the delivery device intended to be used in the clinic should, whereever possible, be used to administer the cell therapy. Where the delivery procedure is novel and this is not practical to perform in the pivotal safety study, due to the small size of the animal model, additional nonclinical studies may be necessary in larger animal models to assess the safety of the delivery device and procedure.

Studies should incorporate traditional safety endpoints such as clinical signs, physical examination, food consumption, body weight measurements, clinical pathology and hematology, organ weights, gross pathology, and histopathology to identify potential targets of toxicity. Where studies are performed on murine animal models, satellite groups may be required for clinical pathology and hematology assessments. Local tolerance and biocompatibility will need to be assessed for novel delivery systems, such as encapsulation materials and novel devices. In addition, the safety of novel routes of delivery will also need to be determined.

Safety pharmacology assessment should be considered on a case-by-case basis depending on the specific characteristics of a cell-based product. Measurements may include assessment of cardiac parameters for heart treatments or behavioral and neural toxicity assessments for cell therapies that target the brain. Reproductive and developmental toxicology will be dependent on the product, clinical indication, and intended clinical population, and typically if required, timing follows the standard regulatory recommendations. Effects on the reproductive system identified in the general toxicity assessments and inappropriate biodistribution, however, may require investigation in more specific studies earlier in development. If reproductive and developmental toxicity studies are required, study designs may need to be altered when an animal disease model is required, and the proposed study designs should be discussed with the regulatory authorities.

8. Conclusions

There has been a rapid advancement in cell therapy development and clinical trials in the last 10 years, with the first therapies licensed and therapeutic possibilities that could revolutionize the treatment of multiple diseases, including the potential for the reversal or removal of certain disease symptoms and conditions. Although there exist commonalities between the nonclinical development paths for cell-based therapies, a standardized approach to nonclinical assessment does not exist—the nonclinical plan is based on scientific rationale using a case-by-case approach. The risk-based approach can provide a framework for developing the nonclinical plan including the assessment of the biodistribution, tumorigenicity, and immunogenicity risks of the product as well as general safety. Innovative study designs, the use of novel in vitro systems, and development of noninvasive imaging modalities may all be required to address specific safety concerns. Given the complexities of the products, however, a pragmatic approach is required, and early and frequent interaction with the regulatory agencies on the proposed designs is recommended. Advances in our understanding of these therapies are rapidly progressing, and as more safety programs are completed and shared within the scientific community, we will continue to develop a more complete understanding of the real risk of each safety issue and strategies to assess them.

References

[1] Jenq RR, van den Brink MR. Allogeneic haematopoietic stem cell transplantation: individualized stem cell and immune therapy of cancer. Nat Rev Cancer 2010;10(3):213–21.

[2] Mason C, Brindley DA, Culme-Seymour EJ, Davie NL. Cell therapy industry: billion dollar global business with unlimited potential. Regen Med 2011;6(3):265–72.

[3] Maude SL, Frey N, Shaw PA, Aplenc R, Barrett DM, Bunin NJ, et al. Chimeric antigen receptor T cells for sustained remissions in leukemia. N Engl J Med 2014;371(16):1507–17.

[4] Lee DW, Kochenderfer JN, Stetler-Stevenson M, Cui YK, Delbrook C, Feldman SA, et al. T cells expressing CD19 chimeric antigen receptors for acute lymphoblastic leukaemia in children and young adults: a phase 1 dose-escalation trial. Lancet 2015;385(9967):517–28.

[5] FDA. Minimal manipulation of human cells, tissues, and cellular and tissue-based products. Draft Guidance for Industry and Food and Drug Administration Staff; December 2014.

[6] Van Wilder P. Advanced therapy medicinal products and exemptions to the regulation 1394/2007: how confident can we be? an exploratory analysis. Front Pharmacol 2012;3(12).

[7] Regulation (Ec) No 1394/2007 of the European Parliament and of the Council: on advanced therapy medicinal products and amending Directive 2001/83/EC and regulation (EC) No 726/2004 November 13, 2007.

[8] Directive 2001/83/Ec of the European Parliament and of the Council of 6 November 2001 on the Community code relating to medicinal products for human use. Off J L November 28, 2004;311:67–128.

[9] Le Blanc K, Pittenger MF. Mesenchymal stem cells: progress toward promise. Cytotherapy 2005;7(1):36–45.

[10] Syed BA, Evans JB. Stem cell therapy market. Nat Rev Drug Discov 2013;12(3):185–6.

[11] Pagliuca FW, Melton DA. How to make a functional beta-cell. Development 2013;140(12):2472–83.

[12] Rezania A, Bruin JE, Xu J, Narayan K, Fox JK, O'Neil JJ, et al. Enrichment of human embryonic stem cell-derived NKX6.1-expressing pancreatic progenitor cells accelerates the maturation of insulin-secreting cells *in vivo*. Stem Cells 2013;31(11):2432–42.

[13] Hao E, Tyrberg B, Itkin-Ansari P, Lakey JR, Geron I, Monosov EZ, et al. Beta-cell differentiation from nonendocrine epithelial cells of the adult human pancreas. Nat Med 2006;12(3):310–6.

[14] Lima MJ, Muir KR, Docherty HM, Drummond R, McGowan NW, Forbes S, et al. Suppression of epithelial-to-mesenchymal transitioning enhances ex vivo reprogramming of human exocrine pancreatic tissue toward functional insulin-producing beta-like cells. Diabetes 2013;62(8):2821–33.

[15] Sharpe ME, Morton D, Rossi A. Nonclinical safety strategies for stem cell therapies. Toxicol Appl Pharmacol 2012;262(3):223–31.

[16] Pessina A, Gribaldo L. The key role of adult stem cells: therapeutic perspectives. Curr Med Res Opin 2006;22(11):2287–300.

[17] Bryder D, Rossi DJ, Weissman IL. Hematopoietic stem cells: the paradigmatic tissue-specific stem cell. Am J Pathol 2006;169(2):338–46.

[18] Keating A. Mesenchymal stromal cells: new directions. Cell Stem Cell 2012;10(6):709–16.

[19] Prockop DJ, Keating A. Relearning the lessons of genomic stability of human cells during expansion in culture: implications for clinical research. Stem Cells 2012;30(6):1051–2.

[20] Ma S, Xie N, Li W, Yuan B, Shi Y, Wang Y. Immunobiology of mesenchymal stem cells. Cell Death Differ 2014;21(2):216–25.

[21] Barkholt L, Flory E, Jekerle V, Lucas-Samuel S, Ahnert P, Bisset L, et al. Risk of tumorigenicity in mesenchymal stromal cell-based therapies–bridging scientific observations and regulatory viewpoints. Cytotherapy 2013;15(7):753–9.

[22] Zangi L, Margalit R, Reich-Zeliger S, Bachar-Lustig E, Beilhack A, Negrin R, et al. Direct imaging of immune rejection and memory induction by allogeneic mesenchymal stromal cells. Stem Cells 2009;27(11):2865–74.

[23] Griffin MD, Ritter T, Mahon BP. Immunological aspects of allogeneic mesenchymal stem cell therapies. Hum Gene Ther 2010;21(12):1641–55.

[24] Reubinoff BE, Pera MF, Fong CY, Trounson A, Bongso A. Embryonic stem cell lines from human blastocysts: somatic differentiation in vitro. Nat Biotechnol 2000;18(4):399–404.

[25] Evans MJ, Kaufman MH. Establishment in culture of pluripotential cells from mouse embryos. Nature 1981;292(5819):154–6.

[26] Martin GR. Isolation of a pluripotent cell line from early mouse embryos cultured in medium conditioned by teratocarcinoma stem cells. Proc Natl Acad Sci USA 1981;78(12):7634–8.

[27] Thomson JA, Itskovitz-Eldor J, Shapiro SS, Waknitz MA, Swiergiel JJ, Marshall VS, et al. Embryonic stem cell lines derived from human blastocysts. Science 1998;282(5391):1145–7.

[28] NIH Human Embryonic Stem Cell Registry: Research using these lines is Eligible for NIH funding. Available from: http://grants.nih.gov/stem_cells/registry/current.html [accessed on 25.02.15].

[29] NIBSC. UK stem cell bank. Available from: http://www.nibsc.org/science_and_research/advanced_therapies/uk_stem_cell_bank.aspx [accessed on 25.02.15].

[30] Puri MC, Nagy A. Concise review: embryonic stem cells versus induced pluripotent stem cells: the game is on. Stem Cells 2012;30(1):10–4.

[31] Takahashi K, Yamanaka S. Induction of pluripotent stem cells from mouse embryonic and adult fibroblast cultures by defined factors. Cell 2006;126(4):663–76.

[32] Gore A, Li Z, Fung HL, Young JE, Agarwal S, Antosiewicz-Bourget J, et al. Somatic coding mutations in human induced pluripotent stem cells. Nature 2011;471(7336):63–7.

[33] Laurent LC, Ulitsky I, Slavin I, Tran H, Schork A, Morey R, et al. Dynamic changes in the copy number of pluripotency and cell proliferation genes in human ESCs and iPSCs during reprogramming and time in culture. Cell Stem Cell 2011;8(1):106–18.

[34] Lebkowski J. GRNOPC1: the world's first embryonic stem cell-derived therapy. Interview with Jane Lebkowski. Regen Med 2011;6(6 Suppl.):11–3.

[35] Schwartz SD, Regillo CD, Lam BL, Eliott D, Rosenfeld PJ, Gregori NZ, et al. Human embryonic stem cell-derived retinal pigment epithelium in patients with age-related macular degeneration and Stargardt's macular dystrophy: follow-up of two open-label phase 1/2 studies. Lancet 2015;385(9967):509–16.

[36] Hency R. Diabetes stem cell Milestone: ViaCyte transplants 1st patient. Available from: http://www.ipscell.com/2014/10/diabetes-milestone-viacyte-transplants-1st-patient-ever-with-es-cell-based-therapy. [accessed on 25.02.15].

[37] Reardon S, Cyranoski D. Japan stem-cell trial stirs envy. Nature 2014;513(7518):287–8.

[38] Getting embryonic stem cell therapy right. Nat Med 2008;14(5):467.

[39] Vierbuchen T, Ostermeier A, Pang ZP, Kokubu Y, Sudhof TC, Wernig M. Direct conversion of fibroblasts to functional neurons by defined factors. Nature 2010;463(7284):1035–41.

[40] Naldini L. Ex vivo gene transfer and correction for cell-based therapies. Nat Rev Genet 2011;12(5):301–15.

[41] Riviere I, Dunbar CE, Sadelain M. Hematopoietic stem cell engineering at a crossroads. Blood 2012;119(5):1107–16.

[42] Perica K, Varela JC, Oelke M, Schneck J. Adoptive T cell immunotherapy for Cancer. Rambam Maimonides Med J 2015;6(1):e0004.

[43] Kanemura H, Go MJ, Shikamura M, Nishishita N, Sakai N, Kamao H, et al. Tumorigenicity studies of induced pluripotent stem cell (iPSC)-derived retinal pigment epithelium (RPE) for the treatment of age-related macular degeneration. PLloS One 2014;9(1):e85336.

[44] Santos E, Pedraz JL, Hernandez RM, Orive G. Therapeutic cell encapsulation: ten steps towards clinical translation. J Control Release 2013;170(1):1–14.

[45] MHRA. Innovation office enquiry form. Available from: http://info.mhra.gov.uk/forms/innovation_form.aspx. [accessed on 25.02.15].

[46] Bailey AM, Balancing tissue and tumor formation in regenerative medicine. Sci Transl Med. 2012;4:147fs28.

[47] Kooijman M, van Meer PJ, Gispen-de Wied CC, Moors EH, Hekkert MP, Schellekens H. The risk-based approach to ATMP development - generally accepted by regulators but infrequently used by companies. Regul Toxicol Pharmacol 2013;67(2):221–5.

[48] Fink Jr DW. FDA regulation of stem cell-based products. Science 2009;324(5935):1662–3.

[49] Cameron BJ, Gerry AB, Dukes J, Harper JV, Kannan V, Bianchi FC, et al. Identification of a Titin-derived HLA-A1-presented peptide as a cross-reactive target for engineered MAGE A3-directed T cells. Sci Transl Med 2013;5(197):197ra03.

[50] Schrepfer S, Deuse T, Reichenspurner H, Fischbein MP, Robbins RC, Pelletier MP. Stem cell transplantation: the lung barrier. Transplant Proc 2007;39(2):573–6.

[51] Maus MV, Grupp SA, Porter DL, June CH. Antibody-modified T cells: CARs take the front seat for hematologic malignancies. Blood 2014;123(17):2625–35.

[52] Koestenbauer S, Zech NH, Juch H, Vanderzwalmen P, Schoonjans L, Dohr G. Embryonic stem cells: similarities and differences between human and murine embryonic stem cells. Am J Reprod Immunol 2006;55(3):169–80.

[53] Prevot A, Semama DS, Justrabo E, Guignard JP, Escousse A, Gouyon JB. Acute cyclosporine A-induced nephrotoxicity: a rabbit model. Pediatr Nephrol 2000;14(5):370–5.

[54] Ito R, Takahashi T, Katano I, Ito M. Current advances in humanized mouse models. Cell Mol Immunol 2012;9(3):208–14.

[55] Tolar J, O'Shaughnessy MJ, Panoskaltsis-Mortari A, McElmurry RT, Bell S, Riddle M, et al. Host factors that impact the biodistribution and persistence of multipotent adult progenitor cells. Blood 2006;107(10):4182–8.

[56] Jin S-Z, Liu B-R, Xu J, Gao F-L, Hu Z-J, Wang X-H, et al. Ex vivo-expanded bone marrow stem cells home to the liver and ameliorate functional recovery in a mouse model of acute hepatic injury. Hepatobiliary Pancreat Dis Int 2012;11(1):66–73.

[57] Ankrum J, Karp JM. Mesenchymal stem cell therapy: two steps forward, one step back. Trends Mol Med 2010;16(5):203–9.

[58] Hoogduijn MJ, Crop MJ, Peeters AM, Korevaar SS, Eijken M, Drabbels JJ, et al. Donor-derived mesenchymal stem cells remain present and functional in the transplanted human heart. Am J Transplant 2009;9(1):222–30.

[59] Fu Y, Kraitchman DL. Stem cell labeling for noninvasive delivery and tracking in cardiovascular regenerative therapy. Expert Rev Cardiovasc Ther 2010;8(8):1149–60.

[60] Brenner W, Aicher A, Eckey T, Massoudi S, Zuhayra M, Koehl U, et al. ^{111}In-labeled CD_{34+} hematopoietic progenitor cells in a rat myocardial infarction model. J Nucl Med 2004;45(3):512–8.

[61] Leech JM, Sharif-Paghaleh E, Maher J, Livieratos L, Lechler RI, Mullen GE, et al. Whole-body imaging of adoptively transferred T cells using magnetic resonance imaging, single photon emission computed tomography and positron emission tomography techniques, with a focus on regulatory T cells. Clin Exp Immunol 2013;172(2):169–77.

[62] Love Z, Wang F, Dennis J, Awadallah A, Salem N, Lin Y, et al. Imaging of mesenchymal stem cell transplant by bioluminescence and PET. J Nucl Med 2007;48(12):2011–20.

[63] Weigert R, Porat-Shliom N, Amornphimoltham P. Imaging cell biology in live animals: ready for prime time. J Cell Biol 2013;201(7):969–79.

[64] Sensebe L, Fleury-Cappellesso S. Biodistribution of mesenchymal stem/stromal cells in a preclinical setting. Stem Cells Int 2013;2013:678063.

[65] Cheng K, Gupta S. Quantitative tools for assessing the fate of xenotransplanted human stem/progenitor cells in chimeric mice. Xenotransplantation 2009;16(3):145–51.

[66] Batzer MA, Deininger PL. A human-specific subfamily of Alu sequences. Genomics 1991;9(3):481–7.

[67] Phillips MI, Tang YL. Genetic modification of stem cells for transplantation. Adv Drug Deliv Rev 2008;60(2):160–72.

[68] Maetzig T, Galla M, Baum C, Schambach A. Gammaretroviral vectors: biology, technology and application. Viruses 2011;3(6):677–713.

[69] Yu SF, von Ruden T, Kantoff PW, Garber C, Seiberg M, Ruther U, et al. Self-inactivating retroviral vectors designed for transfer of whole genes into mammalian cells. Proc Natl Acad Sci USA 1986;83(10):3194–8.

[70] Bear AS, Morgan RA, Cornetta K, June CH, Binder-Scholl G, Dudley ME, et al. Replication-competent retroviruses in gene-modified T cells used in clinical trials: is it time to revise the testing requirements? Mol Ther 2012;20(2):246–9.

[71] Nowrouzi A, Glimm H, von Kalle C, Schmidt M. Retroviral vectors: post entry events and genomic alterations. Viruses 2011;3(5):429–55.

[72] Yang SH, Cheng PH, Sullivan RT, Thomas JW, Chan AW. Lentiviral integration preferences in transgenic mice. Genesis 2008;46(12):711–8.

[73] Zhang L, Thrasher AJ, Gaspar HB. Current progress on gene therapy for primary immunodeficiencies. Gene Ther 2013;20(10):963–9.

[74] Persons DA, Baum C. Solving the problem of gamma-retroviral vectors containing long terminal repeats. Mol Ther 2011;19(2):229–31.

[75] Cavazzana-Calvo M, Payen E, Negre O, Wang G, Hehir K, Fusil F, et al. Transfusion independence and HMGA2 activation after gene therapy of human beta-thalassaemia. Nature 2010;467(7313):318–22.

[76] Corrigan-Curay J, Cohen-Haguenauer O, O'Reilly M, Ross SR, Fan H, Rosenberg N, et al. Challenges in vector and trial design using retroviral vectors for long-term gene correction in hematopoietic stem cell gene therapy. Mol Ther 2012;20(6):1084–94.

[77] EMA/CAT/190186/2012 Reflection paper on management of clinical risks deriving from insertional mutagenesis. April 19, 2013.

[78] Mayshar Y, Ben-David U, Lavon N, Biancotti JC, Yakir B, Clark AT, et al. Identification and classification of chromosomal aberrations in human induced pluripotent stem cells. Cell Stem Cell 2010;7(4):521–31.

[79] Dahl JA, Duggal S, Coulston N, Millar D, Melki J, Shahdadfar A, et al. Genetic and epigenetic instability of human bone marrow mesenchymal stem cells expanded in autologous serum or fetal bovine serum. Int J Dev Biol 2008;52(8):1033–42.

[80] Dhodapkar KM, Feldman D, Matthews P, Radfar S, Pickering R, Turkula S, et al. Natural immunity to pluripotency antigen OCT4 in humans. Proc Natl Acad Sci USA 2010;107(19):8718–23.

[81] Conesa C, Doss MX, Antzelevitch C, Sachinidis A, Sancho J, Carrodeguas JA. Identification of specific pluripotent stem cell death–inducing small molecules by chemical screening. Stem Cell Rev 2012;8(1):116–27.

[82] Lee MO, Moon SH, Jeong HC, Yi JY, Lee TH, Shim SH, et al. Inhibition of pluripotent stem cell-derived teratoma formation by small molecules. Proc Natl Acad Sci USA 2013;110(35):E3281–90.

[83] Thepot A, Desanlis A, Venet E, Thivillier L, Justin V, Morel AP, et al. Assessment of transformed properties *in vitro* and of tumorigenicity *in vivo* in primary keratinocytes cultured for epidermal sheet transplantation. J Skin Cancer 2011;2011. 936546.

[84] Hentze H, Soong PL, Wang ST, Phillips BW, Putti TC, Dunn NR. Teratoma formation by human embryonic stem cells: evaluation of essential parameters for future safety studies. Stem Cell Res 2009;2(3):198–210.

[85] Dressel R. Effects of histocompatibility and host immune responses on the tumorigenicity of pluripotent stem cells. Semin Immunopathol 2011;33(6):573–91.

[86] Ramot Y, Meiron M, Toren A, Steiner M, Nyska A. Safety and biodistribution profile of placental-derived mesenchymal stromal cells (PLX-PAD) following intramuscular delivery. Toxicol Pathol 2009;37(5):606–16.

[87] Park M, Seo JJ. Role of HLA in hematopoietic stem cell transplantation. Bone Marrow Res 2012;2012:680841.

[88] Drukker M. Immunological considerations for cell therapy using human embryonic stem cell derivatives. StemBook 2008.

[89] Swijnenburg RJ, Schrepfer S, Cao F, Pearl JI, Xie X, Connolly AJ, et al. *In vivo* imaging of embryonic stem cells reveals patterns of survival and immune rejection following transplantation. Stem Cells Dev 2008;17(6):1023–9.

[90] Nussbaum J, Minami E, Laflamme MA, Virag JA, Ware CB, Masino A, et al. Transplantation of undifferentiated murine embryonic stem cells in the heart: teratoma formation and immune response. FASEB J 2007;21(7):1345–57.

[91] Pearl JI, Wu JC. The immunogenicity of embryonic stem cells and their differentiated progeny. 2013:37–48.

[92] Badillo AT, Beggs KJ, Javazon EH, Tebbets JC, Flake AW. Murine bone marrow stromal progenitor cells elicit an *in vivo* cellular and humoral alloimmune response. Biol Blood Marrow Transplant 2007;13(4):412–22.

[93] Scheller J, Chalaris A, Schmidt-Arras D, Rose-John S. The pro- and anti-inflammatory properties of the cytokine interleukin-6. Biochim Biophys Acta 2011;1813(5):878–88.

[94] Okamura RM, Lebkowski J, Au M, Priest CA, Denham J, Majumdar AS. Immunological properties of human embryonic stem cell-derived oligodendrocyte progenitor cells. J Neuroimmunol 2007;192(1–2):134–44.

[95] Zhao T, Zhang ZN, Rong Z, Xu Y. Immunogenicity of induced pluripotent stem cells. Nature 2011;474(7350):212–5.

[96] Martin MJ, Muotri A, Gage F, Varki A. Human embryonic stem cells express an immunogenic nonhuman sialic acid. Nat Med 2005;11(2):228–32.

[97] Horwitz EM, Gordon PL, Koo WK, Marx JC, Neel MD, McNall RY, et al. Isolated allogeneic bone marrow-derived mesenchymal cells engraft and stimulate growth in children with osteogenesis imperfecta: implications for cell therapy of bone. Proc Natl Acad Sci USA 2002;99(13):8932–7.

[98] Vaithilingam V, Tuch BE. Islet transplantation and encapsulation: an update on recent developments. Rev Diabet Stud 2011;8(1):51–67.

[99] Casucci M, Hawkins RE, Dotti G, Bondanza A. Overcoming the toxicity hurdles of genetically targeted T cells. Cancer Immunol Immunother 2015;64(1):123–30.

[100] Linette GP, Stadtmauer EA, Maus MV, Rapoport AP, Levine BL, Emery L, et al. Cardio-vascular toxicity and titin cross-reactivity of affinity-enhanced T cells in myeloma and melanoma. Blood 2013;122(6):863–71.
[101] Dominici M, et al. Minimal criteria for defining multipotent mesenchymal stromal cells. The International Society for Cellular Therapy position statement. Cytotherapy 2006; 8(4):315–7.

Glossary

Adaptive immunity The components of the immune system involving cells that are modified to attack specific antigens they encounter.

Allelic Each of two or more alternative forms of a gene that arise by mutation and are found at the same place on a chromosome.

Allogeneic Derived from a different individual and, hence, genetically different from the host.

Allometric Measuring and comparing in relation to the body size/mass of different biological systems.

Amendment A planned and documented permanent change to a GLP study protocol.

Amnionic Derived from the serous fluid in which the embryo and fetus is suspended within the amnion.

Aneuploidy A condition in which the number of chromosomes in the nucleus of a cell is not an exact multiple of the monoploid number of a particular species. An extra or missing chromosome is observed.

Apoptotic A form of cell death in which a programmed sequence of events leads to the elimination of cells without releasing harmful substances into the surrounding area.

Autologous Derived from the same individual to be treated.

Chimera Composed of two genetically distinct components (e.g., cells).

Clinical trial authorization The application to run a clinical trial of a medicinal product must receive authorization from an appropriate regulatory authority.

Clonogenicity The ability of a cell the form clones.

Decellularization The process used in biomedical engineering to isolate the extracellular matrix (ECM) of a tissue by removing its inhabiting cells, leaving an ECM scaffold of the original tissue, which can be used in artificial organ and tissue regeneration.

Deviation An unplanned excursion from the protocol, the impact of the deviation on the study results will need to be assessed and documented. The change may subsequently become a permanent change and a protocol amendment will then be written to permanently change the protocol.

Epigenetic Cellular and physiological traits that are heritable by daughter cells and not caused by changes in the DNA sequence.

Episomal DNA that is extrachromosomal and that may replicate autonomously, e.g., a plasmid.

Etiology The cause, set of causes, or manner of causation of a disease or condition.

Genetic modification The modification of the genetic material of living cells.

Genotoxic Induce damage to the genetic material in the cells through interactions with the DNA sequence and structure.

Homeostatic The tendency to seek and maintain a condition of balance or equilibrium by the constant adjustment of biochemical and physiological pathways.

Human leucocyte-associated antigen (HLA) Highly polymorphic molecule required for antigen presentation encoded within the human major histocompatibility complex.

Innate immunity The component of the immune system that is genetically determined and nonspecific.

Investigational New Drug (IND) Current US Federal law requires that a drug be the subject of an approved marketing application before it can be transported or distributed across state lines. Because a sponsor will want to ship the investigational drug to clinical investigators in many states, it must seek an exemption from this legal requirement. The IND is the means through which the sponsor technically obtains this exemption from the FDA.

ISO 10993 The ISO 10993 set entails a series of standards for evaluating the biocompatibility of medical devices. These documents are part of the international harmonization of the safe use evaluation of medical devices.

ISO 15189:2012 Specifies the requirements for quality and competence in medical laboratories.

ISO/IEC 17025:2005 Specifies the general requirements for the competence to carry out tests and/or calibrations, including sampling. It covers testing and calibration performed using standard methods, nonstandard methods, and laboratory-developed methods.

ISO 9001 The ISO 9000 family is a set of standards dealing with the fundamentals of quality management systems including the eight management principles upon which the family of standards is based.

Major histocompatibility complex (MHC) Genetic locus encompassing a family of highly polymorphic genes encoding proteins that regulate immune responses.

Maximum feasible dose The maximum volume of a specific dosage form that can be administered in an animal via the intended route of administration.

Metastasis The spread of a disease-producing agent (e.g., cancer cells) from the initial or primary site of disease to another part of the body.

Minimally effective dose The smallest dose that will produce a desired outcome.

Multipotent Relating to a stem cell that is capable of differentiating into a limited number of specialized cell types.

Mobilized peripheral blood Mobilization refers to increasing the number of circulating stem cells for collection from a donor's peripheral blood. The donor will receive a blood growth factor called granulocyte colony-stimulating factor (G-CSF), which will be injected once a day until enough stem cells are collected from the blood. This usually takes 4–6 days.

Oligopotent A stem cell that is able to form two or more mature cell types.

Pathophysiological The functional changes associated with or resulting from disease or injury.

Passage Subculturing or splitting cells to keep cells alive and growing under cultured conditions for extended periods of time.

Spatial resolution A term that refers to the number of pixels utilized in construction of a digital image. Images having higher spatial resolution are composed with a greater number of pixels than those of lower spatial resolution.

Sponsors A clinical trial sponsor is an individual, company, institution, or organization that takes responsibility for the initiation, management, and/or financing of a clinical trial.

Telomere A compound structure at the end of a chromosome.

Viral transduction The transfer of genetic material to a cell via a viral vector.

List of Acronyms and Abbreviations

ATMP Advanced therapy medicinal product
BSA Bovine serum albumin
CRO Contract research organization

DNA Deoxyribonucleic acid
EMA European Medicines Agency
EPAR European Public Assessment Records
ESC Embryonic stem cell
FDA Food and Drug Administration
GCP Good Clinical Practice
GFP Green fluorescent protein
GLP Good Laboratory Practice
GMP Good Manufacturing Practice
hESC Human embryonic stem cell
HLA Human leukocyte antigen
HSC Hematopoietic stem cell
ICH International Conference on Harmonization
IHC Immunohistochemistry
IND Investigational new drug
iPSC Induced pluripotent stem cell
IV Intravenous
MCB Master cell bank
MHC Major histocompatibility complex
MHRA Medicines Healthcare Products Regulatory Agency
MSC Mesenchymal stromal cell
NHP Nonhuman primate
OECD Organization for Economic Cooperation and Development
qPCR Quantitative polymerase chain reaction
RCL Replication-competent lentivirus
RCR Replication-competent retrovirus
RNA Ribonucleic acid
RPE Retinal pigmented epithelial cell
SIN Self-inactivating
UK United Kingdom
US United States

Good Manufacturing Practice Compliance in the Manufacture of Cell-Based Medicines

Andy Römhild
Berlin-Brandenburg Center for Regenerative Therapies (BCRT), Charité University Medicine Berlin, Berlin, Germany

Chapter Outline

1.   Outline of the Chapter

The biomedical research aims to improve existing therapy concepts and to develop new options for diseased patients. During the last decade, multiple therapeutic-relevant targets have been identified and the biotechnology sector offered new progressive technologies. This enabled the translation into cutting-edge therapy approaches based on human cells. The production of cell-based medicines (CBMs) requires, in most cases, substantial manipulations of the cell product under aseptic conditions. Therefore, the manufacturing of such advanced therapy medicinal products (ATMPs) is controlled by the Regulation (EC) No. 1394/2007 [1], which defines and classifies ATMP as either gene therapy, tissue engineered, or somatic cell therapy medicinal products [2].

However, there are diverse challenges arising due to the novel and complex nature of these highly innovative drugs. The sterility of the final product, for example, is an issue since cell-based medicines cannot be sterile filtered, autoclaved, irradiated, or sterilized without detrimental effects on the product. Thus, the manufacturing of CBM has to be compliant with the rules of Good Manufacturing Practice (GMP) to ensure highly controlled production conditions leading to a high product quality and a low risk to the patient. Currently, the research, development, and translation of CBM into the clinics are driven by academic institutions, hospitals, and charities with different levels of expertise and experience in the regulatory field [3].

The following passages intend to give a brief introduction into the GMP demands during the manufacturing of CBM. Accordingly, some fundamental regulations and guidelines are delineated. Their implementation along the design of an appropriate GMP-compliant manufacturing environment is outlined, and thought-provoking impulses concerning the production process are given.

2. Quality Management

The content of this section is based on respective regulations and guidelines that provide further specific and profound information [4–11].

2.1 Introduction

Therapeutic cell products are intended to treat patients. In this respect, the manufacturer has to ensure that the product complies with the requirements given by laws and guidelines that should guarantee the patient's safety. Consequently, the production of CBM has to comply with the rules of GMP. The basic idea of GMP is that product quality is reproducibly as a result of a comprehensively designed and controlled manufacturing process. Accordingly, quality cannot be assured as a product of analytical procedures and test methods only.

However, the attainment and sustainment of GMP compliance is not possible without the implementation of a quality management (QM) system. In general, the QM system is a pivotal point that conflates and guides the basic concepts of quality planning, quality control (QC), quality assurance (QA), or the quality risk management, for example. All of these points are interrelated, and they aim to improve procedural flows and to optimize the product consistency. Thus, all have to be considered and none can be outstanding while designing and developing the manufacturing of CBM.

The impact that the QM has or should take will be generally and exemplarily outlined in the following. Initially, some general QM tasks are addressed, while the further subdivision is based on the structure of the GMP guide. Thereby, the interrelation of each single area, respectively the superordinate function of the QM system, should be underlined. Particular requirements of the GMP guidelines, respectively their implementation, is delineated in the corresponding subchapters subsequently.

2.2 QM Objectives

2.2.1 Management Responsibilities

The management (e.g., laboratory supervisor, qualified person [QP]) is responsible for the development, implementation, and efficacy of a QM system. It has to ensure that the definition of "quality" is implemented in a way that requirements and expectations of patients, authorities, employees, suppliers, business partners, and the society are considered adequately.

2.2.2   Quality Policy

The safety of all manufactured products and, in consequence, the patients' safety is central to GMP and its quality policy. Therefore, the analysis, identification, assessment, and avoidance of hazardous activities with the aim to systematically circumvent impairments and mistakes are overriding for the pharmaceutical production of CBM. Consequently, the risk assessment and risk management are essentially important in this respect. In parallel, regulatory requirements have to be fulfilled, and state-of-the-art of science and technology have to be applied. Economy, ecology, as well as occupational safety are also of importance in this respect. Moreover, manufacturing and testing according to specified requirements support the intended consistent product quality. Accordingly, the patient's risk due to inadequate safety, quality, or efficacy is minimized, likewise. It is the management's obligation to reach these quality objectives. This requires an active participation and commitment of all employees involving all suppliers and distributers correspondingly. The QM system is subject to an interactive process of continuous feedback and further improvements. Moreover, it provides the basis for self-inspections and quality audits for regular self-assessments. Accordingly, all areas of the QM system must be provided with competent personnel as well as suitable premises and equipment.

2.2.3   Quality Assurance

The concept of QA covers all aspects that separately or entirely influence the quality of a product. It represents the sum of organized measures that are scheduled to ensure that all products comply with relevant quality criteria that are required for their intended use. Therefore, QA incorporates but is not limited to GMP guidelines.

The management designates a person in charge to implement the QM system. This comprises the circulation of released QM documents as well the termination of expired versions. Moreover, internal revisions and required modifications of QM documents must be guided and executed. The QA manager organizes self-inspections to revise the QM system and has to be actively supported by all employees in this respect.

2.2.4   Quality Control

The QC unit is entrusted with the organization, realization, and documentation of product sampling, testing, and release. Thereby, it ensures that specified relevant tests are satisfactorily carried out before materials or finished products are released for manufacturing or application. The QC department operates independently from all other departments. There is no managerial authority toward the product release decision by the QC manager. QC ensures effective measures and means to reliably control all products that are certified. Accordingly, the validation of all quality control-relevant analytical procedures, as well as the review and approval of in-process control (IPC) procedures during manufacturing, are assigned to the QC unit. Furthermore, it approves specifications, sampling instructions, starting, and packaging materials and delivers the analysis certificate of the finished product.

2.2.5 Quality Risk Management

The systematic process for the assessment, control, communication, and review of risks to the product quality is called Quality Risk Management. Consequently, product safety is the central issue of the risk management approach. It can be applied proactively or retrospectively. Accordingly, quality risks at all levels and grades throughout the entire value-chain are identified and assessed. These risks might be hazardous to the quality and safety of the product. Therefore, they are subjected to quality-assuring measures to accomplish an acceptable risk–benefit balance particularly with respect to the patient.

2.2.6 Change Control Process

Change control processes are focused on managing changes to prevent unintended consequences [12]. Their effectiveness is a key component of the quality system.

As soon as modifications or improvements are subject to implementation, change control processes are initiated, and relevant documents or process sequences are registered and adapted accordingly. This might be caused by reviewed or revised standard operating procedures (SOP), manufacturing or test instructions, as well as altered processes and new equipment, for example. Accordingly, all planned changes with an influence on product quality or other valid statuses (e.g., validation or qualification) are subjected to change control processes. Additionally, they must be assessed and confirmed by competent and qualified employees prior to their realization.

2.2.7 Corrective and Preventive Action

Corrective and preventive action (CAPA) is a regulatory concept that focuses on investigating, understanding, and correcting deviations while attempting to prevent their recurrence [12]. It ensures the timely registration, comprehensive documentation, and adequate handling of all critical and major deviations. The severity of these deviations is classified in accordance to a predefined scheme. Minor deviations are assessed by qualified employees, while critical deviations are reported to the pharmaceutical management, which then is responsible for the evaluation. The implementation and tracing of CAPA is controlled by the QA department and can also be used to prospectively register potential weaknesses and quality risks.

2.2.8 Product Quality Review

Product quality reviews are intended to demonstrate that existing specifications support the process consistency and help to ensure product quality. This periodic trend analysis helps to improve the process management. It is usually done once a year and considers previous review results. In case the initiation of a CAPA process is deemed necessary, its reasons should be documented, and all actions ought to be completed effectively and timely. These procedures should be verified by predefined assessment strategies that are reviewed by the management.

2.3 Personnel

The structure and maintenance of a functional QM system, as well as the manufacture of CBM, are critically dependent on personnel. The organizational structure, qualification, election, and training of the employees are main pillars of an effective GMP-compliant QM system. Having sufficient staff with a fundamental qualification and practical experience is imperative to realizing all duties that are in the responsibility of the manufacturing institution. The areas of accountability that are dedicated to each employee need to be clearly defined. Their extent must not pose quality risks to the product or the worker. The responsibilities are the subject of consistent job descriptions, covering all areas without overlap that otherwise might cause misunderstanding, mistakes, or quality hazards. Personnel in essential positions require appropriate authorization and adequate substitutes. Structure, responsibilities, and absence management are subjects of an organizational chart that provides information for authorities or cooperation partners.

2.4 Premises and Equipment

Premises and equipment must be designed, built, and maintained in a suitable, error-avoiding manner for their intended use. All quality-relevant manufacturing steps are solely carried out with premises and equipment that is easy to clean and maintain. Their quality and layout must avoid impurity and cross-contamination as well as possible. Premises and equipment with an impact on process and product quality are subject to QC actions. This comprises, for example, control labs, manufacturing rooms, stores, and equipment that are named in the corresponding SOPs.

2.5 Documentation

Good documentation is a key element of the QA, respectively the QM system, and a prerequisite for GMP compliance. Created or released documents with relevance to the management of policies, requirements, and process flows within the scope of the QM system are regulated according to a document management system. This system is intended to control, monitor, and record all activities that directly or indirectly influence the product quality. Principally, instructions and records are the two primary types of documentation as defined by the QM system. However, all GMP-relevant documents should follow a controlled life cycle concept that represents the organizational basis for the manufacturing of CBM.

2.6 Manufacturing

The manufacture of CBM must follow defined procedures and specifications to ensure GMP compliance. It is carried out and controlled by qualified personnel. All incoming goods are subject to inspection to guarantee their quality. Impairments and other problems that might influence the product quality are investigated, recorded, and reported to the QA department. Upon receipt, goods and finished products are quarantined in separated stores until they are released for production or distribution. All goods and products have to be stored under controlled conditions to separate different batches.

Finished products as well as starting material have to be protected against (cross-) contamination on each process level. All materials, containers, key equipment, and (if appropriate) premises should be labeled. Labels and batch codes must precisely and clearly indicate all employed container, premises, and equipment. The access to the manufacturing side should be controlled and restricted to trained employees.

2.7 Qualification and Validation

Qualification and validation are elementary parts of a GMP-compliant manufacturing process. The guidelines require, for example, qualified process equipment and validated aseptic handling steps, test methods, or manufacturing processes. Accordingly, the QM has to identify a suitable validation approach (e.g., prospective, retrospective, etc.). Moreover, the scope and extent of these activities have to be determined in parallel to responsible employees or involved departments.

2.8 Self-inspection and Auditing

The QM system guides the schedule of inspections and audits. These are envisioned to monitor the implementation of GMP compliance. Accordingly, internal inspections and external audits of suppliers or contract laboratories are realized. The results are recorded and evaluated against the background of previous assessments with the aim to constantly improve the quality.

3. Documentation

The content of this section is based on respective regulations and guidelines that provide further specific and profound information [7,13].

The implementation of an effective and "living" documentation system is a key component to the GMP-compliant manufacturing of CBM. There should be written instructions for all details with an impact on the product quality. These instructions must be accepted, understood, and supported by the management and personnel. Their accurate realization, including possible deviations, is subject to documentation. Its relevance is significant since authorities and inspectors follow the motto: "Things that are not documented do not exist!"

3.1 General Documentation Requirements

The guidelines specify the types of documents and describe the fundamental basis on which a documentation system is built and maintained. The different requirements for a GMP-compliant documentation system can be summarized as follows:

3.1.1 Generation, Organization, and Control

- Documents must be carefully designed, prepared, controlled, and distributed.
- Approval, signature, and dating are to be done by qualified and authorized employees. The effective date should be defined.

- Documents and their versions have to be uniquely identifiable.
- Copies must be indicated and clearly legible. Copy-related errors have to be avoided.
- The course of review and revision has to be defined by the personnel in charge.
- Training of relevant documents must be recorded.
- The handling of invalid and expired versions must be defined.

3.1.2   Layout and Style

- Documents should not be handwritten, and they should use explicit phrases. They should be well structured and be easy to control.
- In case the entry of data is required, sufficient space should be provided for such entries, and all entries must be unambiguous, clearly legible, and inerasable.
- Corrected entries must be signed and dated.

3.1.3   Content

- Title, type, and intention of documents must be clearly indicated.
- The language should fit their intended use. Instructions, for example, should be written in an imperative, mandatory style.
- Documents within the QM system should be kept up-to-date by more frequent reviews.

3.1.4   Retention

- Records should have a clear relation to the corresponding product batch and its process step. The storage location must be defined, and the record integrity throughout the retention period has to be ensured.
- Batch documentation has specific requirements, especially in case of ATMPs. The directives 2004/23/EC and 2006/86/EC indicate that data required for full traceability shall be kept for a minimum of 30 years after clinical use or expiry [14,15]. The retention period of other documents that are not production related is at least five years.

3.1.5   Document Management System

The documentation management can be paper based or electronically guided. In the latter, the system is subject to a laborious computer validation process. However, the structure of the system can be outlined in different ways. One way to illustrate the structure is exemplified in Figure 1. Documents can be hierarchally grouped. Accordingly, lead documents (level 1) like the Site Master File (SMF) or the Quality Management Handbook (QMH) describe GMP-related activities of the manufacturer and summarize the intended quality philosophy. The second level comprises task- and product-based documents that instruct the responsible personnel. Third-level documents are intended to collect, summarize, record, and report data that are mostly related to product batches. Other documents are likewise quality relevant but with less direct impact on the manufacturing and the finished product. Thus, they are categorized in level 4.

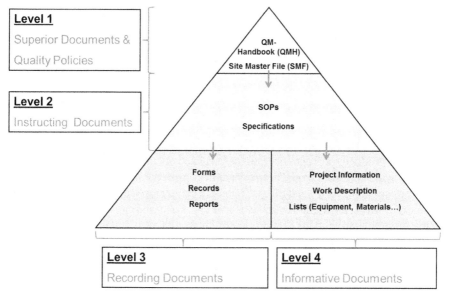

Figure 1 Documentation hierarchy. The hierarchy of different document types is exemplarily structured in four levels according to their intended purpose.

3.2   Definition and Content of Different Document Types

3.2.1   Supervisory Documents

SMF and QMH are higher ranked documents. The SMF briefly describes the manufacturing institution and provides information about the production and control of the manufacturing operations including premises, equipment, personnel, manufacturing processes, and analytical procedures. The QMH specifies the quality policy of the manufacturer in all aspects.

3.2.2   Standard Operating Procedures

Standard operating procedures (SOPs) are documents that contain information as well as general instructions, organizational, administrative, and technical process flows. They are valid for general or specific processes and operations. Supervisory and specific SOPs are concerned likewise. Examples for SOPs are the following:

- Supervisory SOPs
- Manufacturing and packing instructions
- Storage and transport instructions
- Analytical procedures and test instructions
- Operating instructions (e.g., handling, cleaning, maintenance, and calibration)

3.2.3 Roadmaps

Roadmaps contain precise guidance, instructions, and responsibilities for certain operations and tests. They comprise, for example, the following:

- Qualification, validation, and audit plan
- Preventive measure and training plan
- Hygiene and cleaning plan

3.2.4 Records

Records are product-, process-, or equipment-related documents that refer to the corresponding SOPs and roadmaps. They serve to document the operations that are defined in the corresponding SOPs and plans. Thereby, GMP compliance is demonstrated and documented. Records might be simple lists or checklists and can be the subject of validation or qualification plans if appropriate. Their content is usually regarded as raw data. Examples are as follows:

- Manufacturing, test, or monitoring records
- Qualification, validation, or release records
- Cleaning, training, or deviation records

3.2.5 Reports

Results (actual value) of inspections, investigations, and testing are summarized and evaluated in reports. The assessment is based on acceptance criteria that are defined by former surveys (nominal value). Deviations and consequential recommendations are defined, for example, as follows:

- Audit, inspection, or test report
- Qualification, validation, or maintenance report
- Annual reports or risk analysis

3.2.6 Specifications

All essential characteristics of starting materials, packing materials, and finished products that support the identification and are relevant for their release are outlined in specifications. The specification defines the substances to be examined and ensures a clear identification via distinct test methods. It serves as reference and might comprise the test instruction and the respective record.

3.2.7 Risk Analysis

The risk assessment helps to identify, evaluate, and track risks that are related to process flows or equipment, for example. Results might be reported tabular or as continuous text.

3.2.8 Raw Data

Quality relevant data like read offs, calculations, or print outs and electronic data for qualification, validation manufacture, and other assessments are considered raw data.

3.2.9   Log Books

Log books are intended to record quality-relevant actions for all kinds of equipment and important facility parts. Usage, cleaning, sterilization, maintenance, calibration, qualification, validation, changes, repairs, and modifications are the subject of log books.

3.2.10   Register

A register or lists might appear as printed versions or electronic files. They aim to separate operations according to their structure, for example, as follows:

- Equipment, modification, or error list
- Order and storage register

3.3   Standard Operating Procedures

3.3.1   Intention and Purpose

SOPs are intended to facilitate work flows and support the employees in responsible positions. They define tasks, responsibilities, and interfaces. Furthermore, the personnel are instructed to follow standardized procedures that are a prerequisite for a consistent quality. Additionally, SOPs facilitate the batch recording, distribute information equally to all employees, and help to avoid misunderstandings, misinformation, and mistakes.

Consequently, reasons to ignore existing SOPs, for example, too many details, missing clarity, and too extensive descriptions, need to be circumvented. The existence of too many different SOPs or missing support, control, or enforcement by the management are also opposing their efficacy. Accordingly, the authorities judge the disregarding of enforced SOPs to be more critical than missing single instructions as the first points to a systematic failure.

However, SOPs cannot constrain GMP compliance, but they help the manufacturer to fix and sustain valid proven operations. Hence, they support all quality measures, but they cannot replace them.

3.3.2   Structure

The structure of documents in general and SOPs in particular is not predetermined and can be individually designed, although the content has to follow the GMP guidelines. One possible arrangement is exemplified in Box 1.

3.3.3   Life Cycle

The life cycle idea follows central rules to ensure a GMP-compliant document management. Higher ranked documents like QMH, SMF, or other specific documents are prepared with the help of the department, receptively the personnel in charge. The approval is exclusively in the hands of the pharmaceutical management.

Box 1 Structural and Conceptual Aspects of GMP-Compliant Documents

Element/Module	Content
Formal information	Each document must at least contain the following information: Business name/Title/Document type/Document number (ID)/ Version/Pagination This content is intended to appear on each page and is, therefore, best suitable for headlines or footers. However, further points need to be addressed uniquely on one page, such as the following: • Name, date, and signature of the author reviewer and responsible person • Facts comprising scope, historical changes, distribution, and list of attachments
Structural aspects	SOPs, roadmaps, and reports, e.g., should contain the following points: • Cover sheet/Directory/Change history/Intention/Definitions and Abbreviations • Responsibilities/Process description/Applicable documents/ Archiving facts/Attachments
Identification	The document ID should, for example, involve an abbreviation for the document type (like Spec for specification), the document scope, a sequential number, and the version number.

The content of other approval-requiring documents is reconciled already during preparation or before release circulation according to the distribution list. Prior to reconciliation, the author has to sign and date the draft. The review is approved by a second competent person likewise. Preparation, review, and release are dedicated to at least two people, while the author is not allowed to review in parallel.

3.3.4 Preparation

Each employee is allowed to prepare SOPs. However, the management should realize and follow the needs. Accordingly, the author is supposed to be experienced and to know the respective process intimately. Alternatively, intensive reconciliation with users is indispensable.

3.3.5 Review

Similarly, all employees might review documents, although the responsible person should at least be familiar with the respective process. The compiled documents are suggested to be reviewed by local employees for practicability. All interfaces or overlaps to other departments or processes should be handled with caution and are subjected to an approval by the QA management.

3.3.6   Enforcement

The approval procedure of novel and revised documents should be done by different employees, while author and reviewer are supposed to be not identical. The release is dedicated to the responsible specialist, while the QM enforces documents if signatures and training records are appropriate.

3.3.7   Training

Successful and verified training of approved documents is obligatory prior to the release.

3.3.8   Distribution and Administration

Local accessibility of enforced documents must be ensured. The cover letter specifies the distribution. Documents are spread to responsible or significantly affected employees, for example. Signed originals might be filed in the QMH, which should then be indicated in the document's distributor. Copies must be authorized and indicated.

3.3.9   Revision

The approved versions are subject to regular, recorded reviews that are scheduled by the manufacturer. Accordingly, this procedure is devoted to the responsible departments or employees. Major changes should be regulated according to a change control process.

3.3.10   Archiving

The original version of common documents must be archived for more than one year after product expiration or at least five years. However, the batch documentation raised from substantially manipulated CBM has to comply with the ATMP regulation and needs to be retained for 30 years to ensure full traceability. The abrogation of invalid or expired documents is supervised by the QM, receptively the QA manager. These versions must be marked as invalid to avoid a mix-up of valid and invalid documents.

3.4   Raw Data

All records and documents resulting from initial observations or operations related to the manufacture, testing, and release of CBM are specified as raw data. Their definition is certainly more difficult if they are generated by machines or instruments. However, their handling has to be seriously considered as they are part of the documentation system likewise. Consequently, the accurate collection, allocation, usage, and analysis have to comply with GMP standards. Additionally, immediate accessibility, as well as safe and complete archiving, has to be ensured.

3.4.1   Lab Book

Raw data are often recorded in lab books. Therefore, these books should be specific to the job and the employee, bonded, paginated, complete, and readable. Entries can be corrected if necessary. Accordingly, mistakes have to be marked and signed (by the author and reviewer) in a way that the initial entry is still readable. The correction might be briefly explained, but it must always be dated and signed.

4. Qualification and Validation

The content of this section is based on respective regulations and guidelines that provide further specific and profound information [16–22].

The translation of CBM from research to clinical application is inevitably associated with qualification and validation procedures. Both operations are further fragments to ensure consistent quality and prerequisite to any GMP-compliant manufacturing activity. Equipment or systems are the subject of a qualification process while methods are validated. The validation of a manufacturing process, for example, requires the previous successfully completed qualification of all process-relevant equipment including the clean room facility.

The GMP guidelines oblige the manufacturer to identify the specific extent of validation or qualification work that is needed to demonstrate control of critical aspects of the particular operations. Developmental and/or historical data can be employed to identify these critical parameters. Risk assessments are appropriate to determine the scope and extent of qualification and validation operations. Significant modifications of the facility, equipment, and processes are considered to be validated in this respect.

4.1 Strategy/Planning

Planning, execution, and documentation is the order in which qualification and validation work is commonly realized.

The strategic approach should be outlined in a master plan that defines the envisaged aim and respective policy. This document specifies accountabilities, extent, and responsibilities. It further defines the time frame, documentation, and evaluation format. Moreover, it provides a summary of critical steps, analytical methods, equipment, or premises to be used and refers to existing documents. Specifications for successive completion and applied methodology, which might differ depending on requirements and conditions, for example, will be delineated.

4.2 Qualification

According to the GMP guidelines, qualification can be defined as an action of proving that any equipment works correctly and actually leads to the expected results [23]. It should be carried out for a critical system that directly impacts the product quality. Systems with direct product contact, risk of cross-contamination, or systems that are used to record batch data are regarded as critical. This comprises premises, equipment, and technical installations, for example.

4.2.1 Execution

The qualification process is based on the strategic master plan that defines the design and proceedings of qualification plans for specific systems and specific qualification levels. Commonly, the qualification approach is structured as follows:

4.2.2 Risk Analysis

The risk analysis (RA) is intended to identify risk factors and helps to rate and categorize them. These risks should be considered while defining further (premises or equipment-related) requirements. Accordingly, subsequent results are outlined in the following documents:

- User Requirement Specification (URS): All identified requirements should be listed here, but the document should designate those that must be fulfilled to guarantee efficient and consistent processes. The URS is forwarded to the producer of the requested equipment.
- Functional Requirement Specification (FRS): This document contains the producer's suggestion how the specific requirements shall be implemented.

URS and FRS are mainly used if new acquisitions, for example, equipment like centrifuges are envisioned. The risk analysis should be used in any case and especially if already existing equipment has to be qualified. Functional or quality concerns due to the age or cleaning possibilities of the equipment can be managed and documented in this respect.

4.2.3 Design Qualification: "How Should It Look?"

The design qualification (DQ) documents and verifies that the proposed design of the facilities, systems, and equipment is suitable for the intended purpose. It checks the URS against the FRS and revises them referring to GMP, laboratory, manufacturing, and user instructions. In addition, planning directives are controlled for consistency regarding critical process steps and parameters. The results are the subject of a DQ report.

4.2.4 Installation Qualification: "Is Everything in Place?"

According to the GMP guidelines, the installation qualification (IQ) resembles the documented verification that the facilities, systems, and equipment, as installed or modified, comply with the approved design and the manufacturer's recommendations. Accordingly, the goods receipt is inspected for identity, number, and integrity. Moreover, the in-house installation is checked as well as the calibration records, for example. The delivery of user and maintenance documents are as important as the concluding report.

4.2.5 Operational Qualification: "Is Everything Functioning?"

The operational qualification (OQ) represents a documented verification that the facilities, systems, and equipment, as installed or modified, perform as intended throughout the anticipated operating ranges. Therefore, operational parameters like upper and lower limits, power failure, alarm, and safety locking systems are tested. Process data acquisition and transmission are likewise of importance. However, the preparation of training reports and an OQ report must be considered similarly.

All qualification operations can be done under the responsibility of the supplier or manufacturer. Responsibilities, however, should be subject of the supply contract particularly if new equipment is ordered. Typically, these demands are approached before integration into the clean room environment is done. The supplier can be asked or forced to do a factory acceptance test (FAT) to facilitate and accelerate the in-house IQ and OQ actions. This test is done and documented in analogy at the factory side. Upon delivery, only interfaces are tested, which decreases the duration time. The potential detection of errors and deficiencies is likewise of advantage.

4.2.6   Performance Qualification: "Does the Process Run?"

The performance qualification (PQ) is defined as documented verification that the facilities, systems, and equipment, as connected, can perform effectively and reproducibly, based on the approved process method and product specification. Correspondingly, the system should be tested with suitable materials or products including limits and worst-case scenario. Influencing environmental conditions, monitoring data, and trends should be analyzed, respectively. The PQ of ventilation systems, for example, might be extended since it should consider seasonal impacts.

The qualification approach aims to comprehensively ensure the compliance with quality and GMP standards throughout all levels. Successfully completed levels are documented as reports that thereby achieve traceability. This stepwise approach is very interactive, especially if the qualification of existing systems is carried out by the manufacturer. These interactions are depicted in Figure 2, which illustrates that qualification requirements should be driven by process and product-related quality

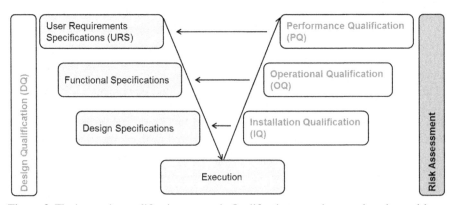

Figure 2 The interactive qualification approach. Qualification procedures are based on a risk assessment. Their realization comprises consecutive steps starting with the design qualification (DQ), which is followed by the installation (IQ), operational (OQ), and performance qualification (PQ). All levels are interrelated and are subjected to feedback, records, and reports.

concerns. This knowledge is further translated to precise requirements that must be successfully implemented and verified. Different functional levels enable feedback loops that could potentially improve the overall outcome.

4.2.7 Requalification: "Is the System Still Compliant?"

The status achieved by qualification must be sustained throughout the complete life cycle of the equipment. This can be achieved by regular calibration and maintenance measures. Requalification is needed if substantial changes or modifications were made. They should be the subject of a change control process, and their severity should be evaluated via risk assessment against the background of potential quality hazards. Nevertheless, a requalification schedule is obligatory to demonstrate (GMP) compliance. The interval can be defined by a time period, manufacturer instructions, or a number of processes, for example. However, the extent should be related to a trend or RA and should be inspired by existing qualification records.

4.3 Supplier Qualification

The qualification of supplier is not a typical qualification activity. Although, it also aims to reduce risks and support a sustained product quality. Furthermore, the workload, for example, during inspections and audits with respect to compliance with internal and external requirements (e.g., specifications and regulations) shall be reduced. Generally, supplier audits are required with respect to the risk management and regulatory demands. These audits or information gathered during supplier inspections can be used to assess their appropriateness. Thereby, at least the following aspects should be considered:

- Supply-chain management and communication
- Quality of the respective material and intended usage during manufacture
- Technological aspects of the purchased material regarding its production and control
- Understanding of GMP requirements and QM system including QC and QA

The manufacturing of CBM products relies on agents and substances that are differently important for the production process. This should be considered while assessing suppliers. The traceability throughout the complete manufacturing of different serum batches, for example, is significantly important and must be ensured by the supply-chain management of the manufacturer. In case of safety recalls or complaints, communication and reporting is central. Technological and QM-related aspects should be assessed with respect to their impact on the manufacture of the CBM product.

4.4 Validation

Qualified premises and equipment are fundamental before initiation of the validation process, regardless if analytical methods or processes are envisaged for validation. Validation studies are inevitable for GMP compliance, particularly if new manufacturing instructions or processing methods shall be introduced. Their applicability in routine operations must be verified according to predefined operations. It must be demonstrated that the defined process or method generates a consistent product or results of adequate quality if defined materials and equipment are used. Results and conclusions have to be documented.

4.4.1   Benefits of Validation

Validation is mandatory according to GMP guidelines and inevitable to demonstrate quality assurance. Nevertheless, it could beneficially impact the process optimization too. The validation process might help to optimize the batch size, enable fewer analyses, and increase the productivity. Moreover, it can reduce costs by avoiding internal and external failures. The first can be related to less rejection, reprocessing, and reanalysis needs, while the latter considers complaints and returns.

4.4.2   Methodology

Qualification and validation can be approached in different ways depending on the given conditions and requirements.

- Prospective validation:
  The prospective approach is done before the manufacturing starts or the method is applied. It has to be demonstrated that predefined specifications are met. Commonly, only three consecutive successful approaches are required and accepted for CBM since the batch numbers, for example, are not comparable to chemical pharmaceutical drugs.
- Retrospective validation:
  The retrospective validation is based on historical data and only acceptable for well-established processes or methods. Recent changes are opposing its usage. Data should originate from 10 to 30 consecutive batches and must be analyzed for consistency. Therefore, batch and packaging records, process control charts, maintenance log books, finished product, storage, and stability data are considered to be reviewed.
- Concurrent validation:
  The concurrent validation approach is carried out in parallel to routine production. It is only accepted in exceptional circumstances (e.g., for orphan drugs or a short shelf life) when availability is more important than the risk. The decision to carry out a concurrent validation must be justified, documented, and approved by authorized personnel.

4.4.3   Validation of Analytical Procedures

Analytical procedures are used to analyze and determine parameters like impurities, sterility, content, phenotype, genomic stability, or potency of CBM. Some more commonly applied methods (e.g., microbiological test) are described in pharmacopoeias. Their application is accepted without validation only if it precisely follows this description. Yet, CBMs are usually manufactured with the help of specific (often supplemented) cell culture media. Therefore, it must be demonstrated (validated) that the cultivation matrix does not negatively affected methods like the test for endotoxins or sterility, even if they are delineated in pharmacopoeias. The analysis or characterization of CBM frequently relies on specific nonpharmacopoeia methods like flow cytometry, which consequently have to be validated. Hence, validation is inevitable for the GMP-compliant manufacturing of CBM. Tests for parameters like identity, impurities, and corresponding limits belong to the most common types of analytical procedures.

4.4.4   Test Parameter

The validation of analytical procedures and methods follows the collective pattern. Initially, a validation team is appointed, the validation concept is scheduled, and the validation plan including protocols is written in accordance to the master plan. Further validation experiments are planned and recorded, and results are reported. Experiments have to address certain parameters to comply with GMP guidelines, and respectively the International Conference on Harmonization (ICH) guideline on the validation of analytical procedures [16].

However, it has to be kept in mind that the primary objective of validating analytical procedures is to demonstrate the suitability of the procedure for its intended purpose. It is the responsibility of the manufacturer, who knows best about the potentially complex nature of the respective product, to choose an appropriate procedure. The parameters described in Box 2 and their corresponding definitions are the subject of consideration in this respect.

Here, it should be again underlined that not all parameters are applicable to all products. Therefore, it is the responsibility of the manufacturer to decide on the design of the validation approach. This stresses the importance of the planning procedure. However, the experiments can be scheduled to enable the analysis of different parameters in parallel. Nonetheless, revalidation might be needed if the drug product synthesis, its composition, or the corresponding analytical procedure is changed.

4.4.5   Validation of Processes

The qualification and validation of analytical procedures should be successfully accomplished before the process validation is envisaged. Process validation comprises all procedures that are related to the manufacturing of CBM. The project proceedings have already been outlined for the validation of analytical procedures and account for the process validation likewise. Conversely, no single experiments, but a complete sequence of consecutive operations are the subject of process validation. The order in which these processes (if subject of the manufacturing) are investigated is not fixed. Nonetheless, it is advisable to initially validate the aseptic processing.

4.4.6   Aseptic Processing (Media Fill)

CBM must be manufactured aseptically due to the inability to sterilize the finished product without detrimental effect on the cell viability. Therefore, the validation of aseptic processing is required to demonstrate product and patient safety prior to routine production. This offers an excellent possibility to train longstanding and to teach new employees, especially if the manufacturing process validation is envisioned thereafter. Hence, each employee should successfully complete at least one media fill before routinely being employed for aseptic processing. Besides, every worker with

Box 2 Definition of Parameter with Relevance for the Validation of Analytical Methods

Parameter	Definition
Specificity	• Ability to assess unequivocally the analyte in the presence of other components like impurities or matrix ingredients
Accuracy (trueness)	• Closeness of agreement between true or accepted and value found
Precision	• Closeness of agreement between a series of measurements
	• The precision should be obtained from multiple samplings of the same homogenous sample across the specified range of the procedure. It is usually expressed as variance, standard deviation, or coefficient of variation and shall be considered at three levels:
Repeatability	• (Intra-assay) precision under the same operating conditions over a short interval of time
Intermediate repeatability	• (Inter-assay) precision within laboratories variations (different days, equipment, etc.)
Reproducibility	• (Inter-operator) precision between laboratories or operators
Detection limit	• Lowest amount of analyte that can be detected in a single sample
	• It can be analyzed by several approaches that might be based on visual inspection, calculation, or extrapolation via standard deviation or calibration curve.
Quantitation limit	• Lowest amount of analyte that can be quantitatively determined
	• It is particularly used for impurities and assessed on spiked samples.
Linearity	• Evidence of proportionality between analyte concentration and result
	• It should be evaluated across the chosen range and in a graphical manner including regression line and its slope.
Range	• Interval between upper and lower analyte concentration in which suitable precision, accuracy, and linearity can be demonstrated
	• The analysis of a finished drug product should normally cover 80–120% of the test concentration.
Robustness	• Measure of capacity to remain unaffected by variations in method parameters
	• It should be already considered during developmental phase and provides an indication of reliability.

responsibility for aseptic processes should at least once a year participate successfully on at least one media fill to demonstrated reliable skills.

The validation of aseptic processing should simulate the routine manufacturing operations as closely as possible using a nutrient growth medium that supports the growing of all kinds of microbiological contaminants. It should demonstrate that the manufacture of sterile products performed with specified equipment, materials, and personnel generates consistently sterile finished products. The approach should include all critical subsequent steps, potential interventions, and worst case situations. Initial media fills usually comprise three consecutive satisfactory runs per shift. The number of filled product units and batches should be comparable or increased compared to the routine process. Additionally, it should be considered that authorities request the processed medium to be completely filled and analyzed. Consequently, the amount of employed medium must be calculated. Accordingly, any significant changes to the equipment, shifts, process, etc. requires a revalidation or repetition. This should be envisaged at least twice a year regardless of changes.

4.4.7 Manufacturing Process Validation

Personnel that are participating in the validation work should be trained appropriately in advance. Process validation is defined as documented verification that the manufacturing approach operated according to its specifications consistently generates a product complying with its predefined quality attributes and release specifications. Certainly, all operations must be carried out by production employees using (qualified) equipment, premises, and materials as applied for routine production. The design of the process validation has to follow the envisaged routine operations. However, the usage of different starting material batches and the manufacturing of product batches in different shifts should be considered if this is scheduled. The process validation is successfully completed if the finished product complies with all predefined release specifications according to validated analytical procedures.

4.4.8 Cleaning Validation

Normally, cleaning procedures are subject of validation to confirm their effectiveness. These procedures are based on microbiological monitoring of equipment, materials, personnel, and clean room facilities, for example, via air sampling, swab tests, final rinse analysis, or contact and settle plates. The cleaning validation aims to verify that specifications for the germ number on surfaces in contact to the product are fulfilled and that a carry-over of microbial contamination is avoided.

The respective validation plan should define the cleaning process including sampling procedures, describe the schedule (who, when, how, which agents, why), and specify the limits based on a risk assessment. Personnel in charge of cleaning must be specifically trained. Three consecutive runs should be realized before validation report and evaluation are undertaken. Revalidation approaches should be carried out periodically and if changes to the drug product or cleaning materials occur.

Nonetheless, cleaning validation might be reduced or not required if only disposables are used for the manufacturing of CBM.

4.4.9 Validation of Computerized Systems

Computerized systems are increasingly used during manufacture, analysis, and release of CBM. In case they are involved in any quality-relevant part of the production process, they need to be validated like all other systems and procedures. The European Medicine

Agency (EMA) and the Food and Drug Administration (FDA), which can be regarded as the EMA counterpart in the United States, provide guidance in this respect [24]. Nonetheless, FDA regulations are not legally binding in Europe. However, the validation of such systems is complicated and should be done by specialized personnel or third parties.

Annex 11 of the European GMP guidelines defines a computerized system as a set of hardware and software components that together fulfill certain functionalities [25]. This comprises, for example, laboratory information management systems (LIMS), electronic signatures, clean room monitoring, and analytical systems. The guideline arrogates the application of a risk management throughout the life cycle of a computerized system. Thereby, patient safety, data integrity, and product quality must be taken into account, and all issues must not be negatively influenced by the system if compared to manual operations. Additionally, personnel qualification, access level, and responsibilities (especially with respect to third parties) must be defined.

The validation approach of computerized systems must follow the GMP guidelines, which means that the systems have to be qualified before the process done by the system is validated. This should cover all relevant steps during the life cycle of the system and includes the manufacturer's justifications of applied standards, protocols, specifications, procedures, and records according to their RA. Accordingly, the following exemplified requirements must be described, implemented, verified, and documented:

- Data integrity during transfer
- Accuracy of critical data
- Security of data storage
- Integrity and security of back-ups
- Audit trail functionality
- Physical and/or logical access restriction

4.4.10   Continued Process Verification

Continued process verification is a recent, more scientifically based idea to approach process validation against the background of improved process consistency and management. Accordingly, it can be summarized as the collection and evaluation of data, ranging from process design to production stage, which enables scientific evidence that a process is capable of consistently delivering quality products. Its basis is a three-staged design. The first stage defines the process design, while the second step is concerned with process qualification to demonstrate reproducible and consistent manufacturing. Stage three finally aims to show continued process verification to underline a constant state of control with regard to routine production.

Continued process verification originates from the pharmaceutical industry and is primarily intended for well-established processes. The scientific basis relies on a statistical evaluation of certain process parameters and, therefore, requires a corresponding product output to ensure product quality. It is questionable if this idea is applicable for CBM, which often follow autologous or personalized approaches with a limited number of products per year. However, this brief introduction will not be further discussed and outlined for these reasons.

5. Premises and Equipment

The content of this section is based on respective regulations and guidelines that provide further specific and profound information [6,26–31].

The manufacturing rooms for the production of CBM and the required equipment have an impact on the product quality. Therefore, they are considered in the guidelines. Both must be sufficiently dimensioned and adequately built to suit the intended purpose while avoiding the risk of product hazards. All product-influencing operations should be carried out with premises and equipment that are easy to clean and maintain. Their design and layout should avoid the risk of cross-contaminations and impurities as well as possible.

Generally, as already outlined, premises and equipment with contact to the product and impact on its quality have to pass the (re-) qualification procedure including the required documentation. The operation of log books is a further requirement to achieve traceability of access, usage, and maintenance. It should be mentioned that both documentations are regularly subject to inspections.

5.1 Premises

The type, size, number, and equipment of all premises in a GMP-compliant facility should be intended, built, and used to minimize mistakes and negative effects on the product quality. This comprises appropriate lightning, temperature, humidity, and ventilation. Moreover, premises should be protected against the entry of unauthorized people, but also against insects and animals that bear contamination risks. The maintenance operations have to be done carefully to avoid product quality hazards. Maintenance activities should be coordinated to avoid multiple production stops.

5.1.1 Design, Layout, and Structure

The design and structure of a facility should support the logical consecutive steps of the production process. This can either be achieved by a physical (e.g., walls, doors) or organizational separation. The latter would intend to carry out activities in the same area but at different time points, for example. Nevertheless, the manufacturers of CBM must follow a clear physical separation of storage, manufacturing, and QC area.

Authorities often assess the suitability and condition of GMP facilities according to the following issues:

- Structural condition and maintenance
- Size based on area and height in relation to its intended use
- Positioning of rooms and their connections to other areas
- Suitability of materials for ceiling, walls, and floors
- Installations and media supply
- Illumination and room ventilation concept
- Implementation of doors and locking systems
- Hygiene concept and status including pest control and occupational safety aspects

5.1.2   Manufacturing Areas

The laboratory and production rooms within the manufacturing area are suggested to be access restricted and only entered by trained personnel. Visitors and inspectors must receive training and should be accompanied by employees if they need or want to enter. The production area should be clearly separated from storage and QC premises. Nevertheless, its structure should enable a clear, logical, sequential realization of all planned process-specific operations. Cleaning activities and (cross-) contamination risks should not be conflicted likewise.

CBMs are usually manufactured aseptically since a terminal sterilization procedure is not compatible with the required cell viability. Accordingly, facilities for the manufacture of CBM can only achieve GMP compliance if specific areas for storage, QC, and sterile production are available.

In Europe, aseptic processing must be done in clean room class A. The corresponding nomenclature in the US is different, although the limits are comparable. Clean room or air cleanliness class A can be achieved with the isolator technology or biological safety cabinets (BSC). However, the surrounding environment has to fulfill additional requirements depending on the device that will be used. The use of BSC requires class B environmental conditions. Class B necessitates an interconnection to clean rooms that achieve air cleanliness class C and D. Table 1 shows the requirements that are related to each of these classes and the respective monitoring. The cleanliness level below D is usually not further specified. Air cleanliness classes are classified during the facility qualification period. They need to be determined at rest (without activities) and in operation (with activities). The clean room classes must be regularly, at least once a year, reclassified and requalified.

All rooms of each class must be connected via personnel and material airlock systems. The transfer of personnel to the next higher level of cleanliness is usually associated with a change of clothes to minimize microbial and particulate carry-over or contamination of the protective clothing. These airlocks should, therefore, clearly indicate where the next class level begins. These can be released with sit-overs, for example, which in parallel simplify the changing procedure. The layout of such airlocks has to consider storage area for the clothing. Moreover, an interlocking system and/or a visual warning system is needed to prevent the opening of more than one door at the same time or before air exchange (flushing) is completed. Flushing is needed since two different cleanliness levels meet in an airlock. An intensive air exchange ensures that the airlock achieves the cleanliness level of the connected clean room with the higher grade before its entry. The manufacture of CBM has to be done with a positive pressure gradient from class A to class D and the outside to ensure that particles will move away from the aseptic area. Thereby, it is intended to protect the product quality. This intention is twisted if the manufacturing of certain gene therapy products is envisaged. In this case, level A with positive pressure must be surrounded by a negative pressure area class B to protect the environment. Accordingly, pressure indicators are needed in any case, and their recording is mandatory to demonstrate compliance with the regulations.

Table 1 Class Limits for Air Cleanliness and Microbiological Contaminations

Class	Air Cleanliness Limits (At Rest/In Operation)		Microbiological Contamination Limits (CFU)			
	≥0.5 μm	≥5 μm	Air Sample (CFU/m³)	Settle Plates	Contact Plates	Glove/Fingerprint
A ISO 5, M3.5, class 100,	3.520/3.520	20/20	<1	<1	<1	<1
B ISO 5, M3.5, class 100,	3.520/352.000	29/2.900	10	5	5	5
C ISO 7, M5.5, class 10000,	352.000/3.520.000	2.900/2.900	100	50	25	n.a.
D ISO 8, M6.5, class 100000	3.520.000/n.a.	29.000/n.a.	200	100	50	n.a.

This overview shows the particle and microbiological contamination limits for each clean room class (A–D). Particle limits are shown for both particle sizes (≥0.5 μm and ≥5 μm) and both operation modes (at rest/in operation). Microbiological contamination limits are depicted as colony forming units (CFU).

The ventilation and filter systems must be designed to enable a pressure difference of 10–15 Pa between each cleanliness class as well as adequate humidity and temperature within the facility. Filtration of supply air is commonly a staged-process that employs a terminal high efficiency particulate air (HEPA) filter to achieve the required separation efficiency. If the system is set up to work only with fresh air, the corresponding intake passage should be kept at a distance to exhaust air systems. Filter exchange, test of tight fit, and integrity are subject to regular maintenance and an essential part of (re-) qualification activities. Since the ventilation system consumes a lot of energy, an automated night reduction system might reduce these costs. Nonetheless, it is subjected to qualification and validation.

Clean room facility design should consider that all exposed surfaces must be smooth, impervious, and unbroken to facilitate cleaning activities. Therefore, simple geometries without sharp edges or corners are helpful, and also to minimize the clean room surface area. The chosen materials should minimize the shedding or accumulation of particles, dirt, or micro-organisms, while in parallel allowing repeated application of cleaning and disinfecting agents without deterioration. Furthermore, design and geometries must support hermetically sealing and barometric pressure stability. System walls and ceilings that are powder coated and rubber floors comply with these features and are often used for that reason. Nevertheless, all joints must be sealed with silicone, and the floor should have a concave molded wall flashing. Furthermore, the floor should be electrically conductive since static charging can otherwise damage electronic equipment.

The number of installations like tables and shelf storage areas should be minimized but still satisfy the needs. Otherwise, cleaning procedures might be impaired, and the place for storage might be misused for unnecessary things that increase the contamination risk. Accordingly, the implementation of incubators, fridges, and other devices might be realized to enable maintenance from the outside. Such service corridors simplify and accelerate the repair or maintenance and can circumvent the entry of technicians and their tools. However, the entry and garment changing can be facilitated if mirrors are placed in airlocks. Their correct gapless installation is a very helpful tool to prove the accurate fit of the garment.

The manufacture of different CBM products or batches is only possible if the facility provides separated areas for their simultaneous production and the risk of cross-contamination is acceptable. Simultaneous work with different products at the same time in the same area is forbidden and may not be justified by additional extensive labeling. Further precautions might be required to minimize this hazardous risk. Consequently, dedicated equipment or production in campaigns might be necessary. However, after production is finished and new products or batches are envisaged, an extensive cleaning of facility and equipment is absolutely crucial. The operating mode of the clean room facility should be visible for all employees during manufacturing. Appropriate systems must ensure that the operating personnel are warned if particle or pressure limits are in danger and might impact product quality. This can be realized via optic and/or acoustic signaling.

5.1.3 Storage Area

Only materials and substances (e.g., stock solutions, cleaning and disinfecting agents, etc.) that are urgently needed for the production are stored within the manufacturing area (class level A–D).

All other materials and items must be retained in other storage areas outside of the production rooms. These areas have to be clearly physically separated not only from the manufacturing part but also from the QC part of the facility. The storage capacity should be sufficiently dimensioned to allow systematic organization of different material and product categories like starting materials, packaging materials, intermediate, bulk, and finished products. Accordingly, the storage area should be well planned, since storing CBM and corresponding materials often requires substantial refrigerator and freezer capacities. In case manufacturing activities are extended to further products, these capacities might become a limiting factor otherwise.

Generally, GMP-compliant storage of CBM must be structured in separated zones for the storage of materials that are under quarantine after receipt, approved for production, and suspended due to expiration or recall, for example. All items should be visibly indicated, for example by red, yellow, and green stickers, according to their corresponding status in this respect. CBM products are often cryopreserved before all relevant tests results (e.g., sterility) are obtained for their final release. This requires additional space for liquid nitrogen tanks, which must be assured for products under quarantine and certified finished products equally. Products should be stored in the gaseous phase to avoid cross-contamination risks. Nevertheless, if the product is manufactured with or contains infectious material, the authorities might ask to store each product physically separated.

Consequently, good storage conditions must be implemented and fulfill the product-specific needs for all materials. This means that stores should be clean, dry, and enable acceptable temperature limits. Furthermore, they must be spacious enough to avoid confusion, cross-contamination, or other quality hazards. Storage conditions that differ from usual room temperature conditions must be monitored and documented to achieve GMP compliance. Documentation can be paper based with daily entries, for example, if no electronic or online monitoring system is available. Resulting data might be used for trend analysis and support scheduling of maintenance or defrosting activities of freezers, for example.

The storage area, particularly quarantine and approved product stores, should be access-restricted to avoid the entrance of unauthorized personnel.

5.1.4 Quality Control Area

Indeed, the QC area should be separated from other GMP-relevant areas. Moreover, control laboratories of biologics, microbiologics, and radioisotopes should also be separated from each other. Nevertheless, all of them have to be suitable and provide enough space to carry out intended operations without detrimental impact on the product quality. This is particularly important if microbiological testing is done in-house and not outsourced.

5.1.5   Ancillary Area

The ancillary area might comprise, for example, an archive, office, sterilization room, cleaning storeroom, maintenance corridor, changing room, and a refreshment area.

Documentation of CBM batch records has usually to be retained for 30 years, which has to be considered if archiving is envisaged. Moreover, the archive has to be lockable, comply with fire prevention regulations, and ensure appropriate controlled (e.g., humidity) conditions.

Office space should not be underestimated since GMP compliance is largely achieved by documentation work. Similarly, changing rooms should be reasonably designed and in accordance with the number of employees. Additionally, wearing professional garments supports the reduction of the overall particle burden inside the quality-relevant facility areas. The work in nonbreathable clean room garments under controlled conditions including low humidity and high air exchange rate is hard and exhausting. Therefore, the design of social and refreshment areas must be considered not just for aspects of labor law.

5.2   Equipment

The equipment must be designed, located, and maintained to suit its intended purpose. It should be easy to clean to avoid potential contaminations. Additionally, proper calibration is needed to ensure consistent accurate performance and to enable qualification and/or validation procedures. All equipment must comply with its specifications, irrespective of whether it is used to generate data or maintain GMP standard conditions. In case equipment has to be repaired or maintained, records should be kept in dedicated log books. Trend analysis of equipment is a powerful tool to identify malfunctions as early as possible (e.g., refrigerators as already outlined). Therefore, the review interval should be assigned to assure this. Equipment can either be the subject of preventive or curative maintenance, which can be summarized as follows:

- Preventive maintenance:
  Single items are regularly exchanged without an occurrence of malfunctions. This can be scheduled according to trend analysis, experience, or life expectancy. It is particularly used for instruments or parts of systems that have no back-up and helps to reduce the risk of serious break-downs.
- Corrective maintenance:
  Corrective maintenance aims to repair equipment that is defective and already malfunctioning. These parts might be subjected to preventive maintenance in the future. If this is not possible due to the nature of the system (e.g., for computerized systems), alternative solutions should be planned. Systems might be duplicated or spare parts are retained in-house.
  Accordingly, equipment that is defective, subjected to maintenance activities, expired, or not used should be indicated to avoid its false usage for manufacturing or QC procedures. Furthermore, items with product contact must not interact in a way that the product quality is hampered or other risks arise. Some items of the most commonly employed equipment for the manufacture of CBM that should be subjected to planned maintenance and (re-) qualification are listed in the following. All of them are relevant for the product quality, and the operation within their specifications must be checked regularly and consistently documented throughout the production phase.

5.2.1   Biological Safety Cabinet

Biological Safety Cabinets (BSC) are also known as laminar air flow work stations (laminar flow) or clean benches. Laminar flows are inevitable for the manufacturing of CBM since all product manipulations have to be done aseptically. Off course closed systems can be used to reduce these needs, but currently, production steps are often too diverse to be realized in such systems. BSC provide environmental, personnel, and/ or product protection against hazardous particles. According to their safety function, they are classified as level 1, 2, or 3 systems. The name already implies that their functionality is based on a laminar air flow inside the cabinet. This should be kept in mind during production activities, because rapid movements and an overload with materials disturbs the unidirectional laminar flow and induces turbulences. Turbulent flows are uncontrolled to some extent and endanger the product sterility or quality. BSC should enable a homogenous air speed in a range of 0.36–0.54 m/s at working position. Air speed, laminar flow, filter tests, and changes should be similarly subject to maintenance and (re-) qualification.

5.2.2   Incubator

Substantial manipulations of CBM often comprise extensive in vitro cultivation or expansion periods that are carried out in incubators to mimic physiological conditions. Therefore, atmospheric conditions inside the incubator must be controlled and documented. Important parameters are temperature, humidity, and gas content (e.g., carbon dioxide and/or oxygen) in this respect. Humidification is achieved by water evaporation inside. Accordingly, only sterile water without any harmful supplements should be used. The inside of incubators can be composed of multiple single chambers to ensure physical separation of different product batches, for example. Consequently, calibration and (re-) qualification activities should consider homogenous conditions in each chamber. This can be realized by mapping approaches with multiple measuring instruments and points. After completion of each manufacturing approach, the incubator must be carefully disinfected. An additional decontamination procedure should be scheduled and validated.

5.2.3   Centrifuge

Many CBM products use certain subpopulations of the peripheral blood as a source of starting material. Initially, these cells are frequently isolated by density gradient centrifugation. Even if the manufacturing process is based on processing of whole blood, the cell harvest or other enrichment procedures rely on centrifugation steps, likewise. The installation and operation of these machines must be carefully checked and monitored since they might generate a high particle burden. This can be circumvented by centrifuges with clean room certifications. If these expensive apparatuses are needed, they should be the subject of cautious considerations (e.g., costs versus particle burden or cleanliness classes). Centrifuges are available for diverse g-force limits, with or without cooling system and with diverse rotors and inserts, which in turn can impact the achievable g-force. Manipulations of CBM products, especially during isolation procedures, rely on reduced or enhanced cell metabolism, which can be influenced by temperature, for example. Temperature reduction might also be required for cell harvesting to prepare subsequent freezing steps and to support the product viability. Similarly, all aforementioned points should be considered with respect to potential future process changes or new manufacturing approaches and techniques.

The overall functionality is essential since only the interplay of g-force, temperature, and brake ensures a correct isolation of cells via density centrifugation, for example. Therefore, all features with relevance to the manufacturing process are supposed to be subject of maintenance and (re-) qualification testing.

5.2.4 Refrigerator and Freezer

The manufacturing of CBM is often based on extensive in vitro cultivation periods with systematic feeding strategies. Cell culture media, stock solutions, or other supplements like cytokines are stored in refrigerators or freezers meanwhile to avoid their destruction. Some of these agents might be sensitive to freezing and thawing, which implements that a homogenous temperature distribution inside the machine is crucial to their quality and efficacy. Accordingly, calibration and temperature mapping during maintenance and (re-) qualification is mandatory. Obviously, machines for long-term storage of materials have to be located in the separated storage area of the facility. Nonetheless, dedicated materials (e.g., specific medium) that are used during the manufacturing process might necessitate additional (small) fridges or freezer inside the clean room area. The capacity of these machines has to be well planned, because they are usually integrated in the clean room walls to reduce particle burden, and changes are difficult in this concern.

5.2.5 Particle Counter

Particle monitoring is an essential part of GMP-compliant product documentation, and the corresponding limits are clearly defined in the guidelines (see Table 1). Production activities or other critical process steps that impact the product quality must be accompanied by particle counting in cleanliness class A and B. However, the requirements regarding frequency and sample size of each class differ with respect to their relevance for the product quality. The respective machine should enable appropriate data documentation. Printouts need to be signed and dated to ensure the assignment to the manufacturing step and batch. Caution is needed if the machine employs thermal paper. In this case, printouts should be additionally copied (dated and signed) due to bleaching effects. Generally, particle counters with short sample tubing are preferable due to potential precipitation of particles inside the tubing. Nevertheless, a comfortable placing without impairments during production activities must be ensured if portable machines are employed.

5.2.6 Pipettes

Cell culture work commonly employs pipettes with different volumes. Especially micropipettes with a range of 1–1000 µl are crucial since they are used to pipette highly concentrated media supplements or other stock solutions. Cells are often sensitive to these agents and accept only small concentration ranges. Depending on the process, potential deviating concentrations might accumulated and negatively influence cell expansion, viability, functionality, or purity, for example. This might even cause manufacturing failure due to products that are out of specification. Consequently, (micro) pipettes should be thoroughly cleaned to avoid cross-contamination, but more importantly, they must be systematically maintained and calibrated. Maintenance and calibration is particularly obligatory if pipettes are used for QC procedures, since many methods are often critically reliant on precise volumes.

6. Personnel and Hygiene

The content of this section is based on respective regulations and guidelines that provide further specific and profound information [5,17,32–34].

The establishment and maintenance of a GMP-compliant system ensures the correct manufacture of CBM products and assures a consistent quality of these complex approaches. The executing employees are perhaps the most important part in this QM-based system. Accordingly, the personnel throughout all hierarchical levels and all departments have to be devoted to the principals of GMP and internalize its requirements. Therefore, the manufacturer should have personnel with necessary qualifications and practical experience. Nonetheless, the number of employees must be sufficient to enable each individual to adequately realize their duties. Work overload represents a serious concern to the system in general and the product quality in particular. The GMP regulations provide a broad guidance with regard to personnel requirements. Some universal elements can be easily implemented, but other facets need a more detailed interpretation due to the specific requirements of CBM products. The hygiene, for example, must be particularly considered in multiple aspects since the manufacturing process of CBM is carried out aseptically (for several days or weeks) and a terminal sterilization is impossible. Hygiene becomes even more central if the finished product cannot be cryopreserved and has to be applied before final microbiological test results are available.

6.1 Personnel

The duties assigned to individual employees must be clearly defined. All specific tasks of responsible personnel have to be part of written job descriptions. These key personnel should be authorized to realize their responsibilities without restrictions. Nonetheless, duties can also be carried out by designated deputies if they are appropriately qualified. Key personnel should be clearly dedicated to their area and sector of responsibility. Gaps of responsibilities must be avoided, and overlaps have to be justified. However, overlaps of manufacturing and QC duties are not allowed.

6.1.1 Job Descriptions

The requirements of workplaces should be fixed in written job descriptions. Usually, their content is specified by personnel of the next higher hierarchical level to ensure demands that are close to reality. These descriptions define the desired qualification and experience grade. They further list a comprehensive overview of the aim and duties that are related to the work. Moreover, responsibilities, the authority to issue directives, and the respective supervisor are named. Accordingly, job descriptions also define deputy arrangements. The language used for writing should be precise and understandable.

6.1.2 Organization Chart

The organization chart or organigram is required by the GMP guidelines. It should clearly visualize the organizational structure of the manufacturer. The chart should give an overview of all departments including management and QP. Typically, this comprises the production unit, QC, QA, and/or QM. The organizational chart should name the key personnel in responsible positions according to their department affiliation. Particularly, department manager and their substitutes have to be indicated. Moreover, the QC must be clearly separated from the production unit. Personnel overlaps are not allowed in this respect. In contrast personnel unions between QM and QC or production and QM are allowed.

6.1.3 Key Personnel

Key personnel should be engaged in full-time positions. In Europe, these responsible positions include the head of production, the head of quality control, and the QP. The latter is not known in FDA regulations. Instead, a competent person needs to be defined who is trained to the regulations and is in charge to guarantee GMP compliance. Such duties can be realized, for example, by a quality manager.

The heads of production and quality control, as well as their departments, must be separated and independent from each other. Nevertheless, both generally have some shared or jointly discharged quality-related duties.

Generally, all key personnel are involved in interdivisional activities, like self-inspections. Their common duties also include the preparation of department-specific SOPs, specifications, as well as the avoidance, evaluation, and control of deviations or changes. The qualification, expertise, and experience of the key personnel should be documented. Although precise specifications of their education are rare if described at all.

- Qualified person

   The QP has to ensure that each product batch complies with the respective regulations and, if necessary, the market approval. Different national legislations might also be considered in this regard. GMP-compliant manufacturing, control, and storage are equally important as the usage of approved raw and auxiliary materials, packaging, or labeling. Some of these issues require qualification and validation procedures that should also be agreed to by the QP. However, all of the aforementioned issues have to be approved, before the finished product is registered and released for application. Furthermore, the QP has to participate in self-inspections, authority audits, and the corresponding follow-ups. Accordingly, deviations or changes have to be evaluated, corrective measures must be authorized, and their implementation supervised. The communication and exchange of information with authorities is also under the responsibility of the QP.

Nevertheless, the guidelines or directives do not precisely specify the requirements that a QP for the manufacturing of CBM has to fulfill. General expectations are, for example, completed university studies or recognized equivalents that were extended over a period of at least four years. These should include theoretical and practical courses in a scientific discipline like biology, medicine, veterinary medicine, biochemistry, or biotechnology. Moreover, candidates shall be practically experienced in QC and IPC for CBM product and their respective facets like general cell therapy, gene therapy, and tissue engineering. Overall, due to the requirements related to the manufacturing, control, and application of CBM products, a broad knowledge of pharmaceutical, medicinal, developmental, and clinical trial-related questions is desirable. This is especially valid if the CBM product is not approved, but rather under development and considered as an investigational medicinal product.

- Head of production

The main responsibility of the production manager is to ensure that the manufacturing process complies with predefined written instructions and GMP regulations.

Consequently, the head of production has to ensure that the production and storage of products is in accordance with the specific documentation to achieve an adequate and consistent quality. Accordingly, the responsibility for the qualification of premises and equipment, validation of critical manufacturing steps, and the implementation of production-related instructions has to be taken. Additionally, it must be ensured that production records are written, evaluated, and signed before they are handed over to the QC department. The appropriate maintenance of premises, equipment, and department-specific duties like the review and/or updating of manufacturing-related documents are also under the responsibility of the production manager. Moreover, this person has to approve and monitor suppliers. Besides, the head of production is in charge to ensure that all employees of the department receive initial and continued training according to their dedicated responsibilities and adapted to the (latest) needs.

The educational requirements of the production manager are not defined at least by international guidelines. Nonetheless, it is expected that the person is qualified and experienced enough to adequately realize all responsibilities. The qualification standard is strongly dependent on the area and extent of responsibility. Accordingly, the type and number of products, complexity of manufacturing equipment, as well as number and qualification level of the employees should be considered. An academic education might be desirable although it is not binding. However, the respective person can only be employed according to their qualification and knowledge.

- Head of quality control

The QC manager is in charge to approve or reject all materials and products including starting and packaging materials or bulk and finished products. This has to be done according to predefined instructions and GMP regulations. The head of QC is likewise responsible to ensure the realization of all scheduled tests and process controls including their corresponding validations. Accordingly, specifications, instructions, and test methods have to be approved. QC management also includes the evaluation of batch records and department-specific maintenance activities including relevant premises and equipment. Responsibilities regarding the training of QC personnel are comparable to those of the production manager.

However, the head of QC is uniquely in authority to assign certain quality-relevant duties or controls to third parties or contract laboratories. Although, all routine testing like IPCs or finished product controls must be done in-house. Only specific analytical procedures that require suitable equipment and personnel can be entrusted to contract labs. These procedures include, for example, sterility, endotoxin, or *mycoplasma* tests. Nevertheless, the manufacturer, respectively the head of QC, and the QP are responsible to guarantee a GMP-compliant product quality. Consequently, it must be assured that the contract lab has suitable premises and equipment to ensure that tests are carried out according to state-of-the-art techniques.

The qualification background of the QC manager is comparable to those of the production manager since all general requirements are equally important.

- Head of quality assurance
- The QA department is not mandatory, and also, a description of QA-specific responsibilities is hard to extract from the guidelines. Nonetheless, the QA manager is supposed to ensure the implementation of GMP regulations and other laws or guidelines with relevance to the manufacture of CBM products. Further, classical QA duties comprise the realization of self-inspections, maintenance of the quality system, as well as review and revision tasks.
- Overall, these responsibilities are very close to general QM liabilities. Qualification requirements are not described, although it can be anticipated that at least substantial experienced should be mandatory due to the diverse obligations.

6.1.4 Shared Responsibilities

Although the departments of production and QC must be clearly separated, even if only a small number of people is employed, there are some shared responsibilities. This is, for example, caused by duties that are dedicated to QC but require precise knowledge of the product relevant requirements or characteristics. Some of these shared duties are listed in the following:

- Definition and control of storage conditions regarding materials and products
- Sampling, investigation, and control of materials or factors related to product quality
- Approval and control of suppliers and contractors
- Definition of product and release specifications, appropriate test methods, and their validation
- Documentation storage and control of GMP compliance
- Working hygiene and personnel training.

6.2 Personnel Training

The manufacturer should provide adequate and continued training for all personnel that might affect product quality. This includes employees for production, QC, packing, storing, technique, cleaning, purchase, research, and development. Moreover, visitors or inspectors should be similarly trained for hygiene and the use of specific (protective) garments. Two aspects of central importance concern the appropriate documentation of training and the required evidence of qualification (e.g., certificates) from the trainer. Training should intend to convey contents and their implementation. It should achieve GMP understanding, qualification, motivation, and last but not least, product safety.

6.2.1 Training Concept

The development of a training concept should be based on identified needs and aim to satisfy them. Training should not be understood as an end in itself since it requires valuable resources (e.g., trainer, working time, money). The final concept has to answer the following questions:

- Why is this training necessary?
- Who must be trained?
- Who is training?
- How often will they be trained?
- How long will they be trained?
- What will be trained?

6.2.2 Training Reasons

New employees must receive an initial theoretical and practical training. Basics about working safety and regulations of GMP as well as hazardous materials should be concerned due to the demands while manufacturing CBM products. Afterward, newly employed personnel should be trained according to their assigned responsibilities that are defined in their job descriptions. The final goal implies that each employee is able to autonomously execute all duties and fulfill all quality requirements at any time. Moreover, repetitive, continued, or advanced training can be scheduled. Further needs might originate from novel or revised GMP documents (e.g., SOPs, specifications, or official guidelines) and inspections. Training is often appointed prior to inspections or if objections require retraining.

Another central aspect of training concerns company or working hygiene. These aspects cover behavioral rules related to the potentially infectious nature of CBM products or their starting materials and the personnel safety in general. In particular, employees designated to aseptic procedures must receive specific and advanced training concerning any sterile handling, adequate behavior, and clothing, for example.

6.2.3 Training Methods

The training methods should consider the human nature. This implies that most people learn best if they talk about the subject of training or if they practice it. Therefore, training sessions will be more successful if they are well structured and provide convincing arguments, comparisons, or examples. Accordingly, talks supposed to be given in a nice atmosphere use short but precise and simple words. Besides convincing concepts, the methodical concepts should initially issue simple things before complex points are discussed. Moreover, employees should be contributive and involved in decisions.

Training might be carried out as read and understood session or by external trainers. Additionally, pier training can be offered and is, for example, often used for new methods or media fills. Thereby, the employee needs to successfully accomplish a predefined number of runs to acquire the respective qualification.

6.2.4 Training Topics

Relevant training topics can be related to basic GMP issues that are elementary to all employees. This includes, for example, the following:

- GMP philosophy and its meaning to the manufacturer of CBM
- All aspects of hygiene like protective clothing, make-up, jewelry, etc.
- GMP-compliant documentation (e.g., corrections, labeling, etc.)
- Clean room classes and specific garments
- Qualification and validation.

On the other hand, subjects of training can also be very specifically linked to single operations or processes within the production department, for example. Some potential topics are listed in the following:

- Equipment instructions and data recording
- Environmental or production accompanying monitoring
- IPCs, cleaning, and disinfection
- Training for cleaning personnel and maintenance technicians.

6.2.5 Training Records

The documentation of training sessions is almost equally important as the training itself. Accordingly, inspectors articulate the view that training was not passed if there is no other evidence for its realization. Therefore, employees might be considered as not adequate and seen as a potential risk to the product quality. Hence, training must be documented. This can be done, for example, via attendance lists, certificates, or register in personnel files. The record should provide information about the following:

- Date, topic, trainer, and participants
- Reason and responsible person for training

Training records should be signed by student and trainer. Personnel training is a popular subject of inspections and should, therefore, be comprehensively documented.

6.2.6 Training Evaluations

The monitoring of successful training can, for example, be approached with concluding discussions, assessments by the trainer, oral testing, and personal or anonymized questionnaires. Other indicators might monitor the implementation of training contents or the reduction of deviations that were the subject of training. Nevertheless, if training efficacy is evaluated, certain success parameters must be defined. These parameters should be in line with the demands and not be too soft. Similarly, the grade of compliance with these parameters should be monitored and used to determine further training needs.

6.2.7 Training of Aseptic Processing

Regular continued training is fundamental to all personnel employed in aseptic processing. The respective training has to cover diverse topics, must be well structured and should include issues such as the following:

- Basics of clean room-related hygiene, garments, behavior, and equipment techniques
- Cross-contamination, environmental monitoring, and cleaning concepts
- Practical sessions and success control (media fill)

6.3 Hygiene

The content of this section is based on respective regulations and guidelines that provide further specific and profound information.

Hygiene is of general importance for the manufacturing of medicinal products in order to ensure a consistent quality and GMP compliance. Nonetheless, cleanliness and sterility becomes an essential key element if these products are based on cells and their production process relies on aseptic processing. The fact that finished CBM products cannot be sterilized and are frequently applied with advance release procedures further underlines this significance. However, hygiene covers personnel and manufacturing related issues. The hygiene concept should aim to equally implement personnel as well as production aspects and to intertwine them. It must be promoted by the management and extensively discussed during training conferences. The intention has to be focused on sustained product quality and patient safety of CBM. Envisaged procedures must be clearly understood and strictly followed by all employees.

6.3.1 Personnel Hygiene

All personnel should be subjected to regular medical examinations, especially upon recruitment. Checkups, particularly for production employees, have to consider potential threats to the product by the personnel but, likewise, potential risks for the personnel related to the handling of the product. Infectious diseases like hepatitis or the human immunodeficiency virus should be named in this respect. This implies that the doctor is informed about the specific (GMP) requirements regarding the examination. Vaccination should be considered in this respect if potentially infectious materials (e.g., blood) have to be handled during the manufacturing of the respective CBM product.

GMP guidelines presume that it is the manufacturer's responsibility to be informed about health conditions of the personnel that might impact product quality. Accordingly, employees suffering from diseases (e.g., open lesions, infections, etc.) that might affect product quality should be excluded from respective activities.

The personnel must be aware of potential contamination risks that finally might impact the product quality. Thereby, employees themselves represent a major source of contamination. Accordingly, hand hygiene is a central issue in clean room facilities. All personnel should be instructed to thoroughly disinfect their hands upon arrival at the facility to avoid the carry through of germs from the outside. The correct disinfection of the palms is not trivial and should be subject of theoretical and practical training sessions to guarantee successful execution.

6.3.2 Garments

Professional garments are used for different purposes. They have a protective function toward the contamination of the product but also to prevent hazards of the personnel. In addition, they might be used to indicate the affiliation of employees with respect to their responsibilities.

The structure of GMP facilities for the manufacture of CBM is usually composed of an encapsulated clean room area ranging from cleanliness class A–D and a surrounding area. This area is often not further specified and comprises storage, QC, archive, and an office area, for example. Upon entering the surrounding GMP area, garments should be removed to avoid dirt that might come close to the clean room entries later on. Of course, this is to some extent related to the arrangement or connection of these areas. Nevertheless, at least outdoor shoes should be changed. It might also be suitable to change street wear for professional clothing. However, lab coats should be the standard clothing while working in QC or storage areas. Additionally, gloves can be appropriate to avoid contaminations and to protect the personnel. Vice versa, it is obligatory that this garment must be deposited before leaving the facility, entering the office, or rest rooms, for example.

Aseptic processing necessitates specific garments that employees are gowning while entering the clean room area. Clean room clothing must fulfill specific requirements with regard to the intended use. The material should restrain particle emission and be resistant to abrasion due to the release of fibers. Moreover, electrostatic and carrying comforts have to be considered. The ease of decontamination and sterilization should go along with its protective function. Usually, clean room garments are sterilized and delivered in sterile packing.

The clothing concept can be diverse, for example, with respect to the avoidance of cross-contaminations and might be based on results obtained from microbiological monitoring. However, a potential arrangement of basic requirements is exemplified according to the route of entry in Table 2. The dressing procedure can be summarized as follows:

- The employee has to wear professional clothes, a hairnet, and face mask before entering clean room class D, which usually is a locking room. Additionally, a beard mask might be required to cover the face.
- In cleanliness class D, the personnel should undress and change the shoes. Before proceeding to the next locking room, hands must be washed.
- The locking room from class D to C must be equipped with sterilized underclothing, shoes, and gloves. The employee has to change the slippers and put on the respective clothes, including socks and gloves. Hands, respectively gloves, should be disinfected before entering clean room class C.
- Clean rooms of cleanliness class C are commonly built as preparation or storage rooms. However, a locking room leading from class C to B is mandatory. In this room, the personnel have to put on an additional overall that covers the complete body surface including the

Table 2 Dressing Procedures for Clean Room Garments

| Cleanliness Class | Item(s) of Clothing | | Remarks |
	Before Entry	After Exit	
n.a./D	Professional clothes or lab coat, hairnet, face mask, beard mask (if required)	Underwear, new slippers	Garment is changed; hands should be cleaned
D/C	Underwear, shoes	Sterilized underclothing incl. socks, new slippers, gloves	Shoes are changed; underwear is kept; other garment is added; gloves disinfected
C/B	Sterilized underclothing, shoes, gloves	Sterilized overall incl. overshoes and headwear; gloves	Garment is added; gloves disinfected
A	Employees should wear sterilized underclothing incl. socks, hairnet, face mask (beard mask if required), gloves, and new slippers. This clothing layer should be covered by a sterilized overall, headwear, overshoes, and a second pair of gloves.		

This overview exemplifies how garment changes might be scheduled while entering a clean room area.

head. Overshoes have to be used to cover the slippers. A second pair of gloves should confer additional safety in case of leaking. The employee should again disinfect the gloves before entering class B.

All changing and washing or disinfecting activities must follow written instructions to minimize the risk of contaminations in the clean room area. The personnel should receive theoretical and practical training on these topics.

6.3.3 Behavior

Smoking, eating, drinking, chewing, storage of food, and personal medication must be restricted to designated areas that are separated from product quality-relevant areas like stores, QC labs, or manufacturing rooms. Regarding employees assigned to aseptic processing, it should be considered that particle emission is elevated up to several hours after smoking.

Personnel involved in the manufacturing of sterile preparations are not allowed to wear wristwatches, make-up, or jewelry in clean areas due to an increased risk of contamination. These employees should be instructed to report any personal or familiar health condition that could be detrimental to the product quality. Generally, only a minimum number of personnel should be present during aseptic processing steps, and their movements must be restricted and controlled to avoid the needless shedding of particles.

6.3.4 Production Hygiene

Personal hygiene and GMP-compliant conscious behavior is prerequisite to the pursuing concept of production hygiene. Consequently, this concept involves premises, equipment, and materials. It requires intensive continued training that should provide information about microbiological aspects and the corresponding contaminations risks for CBM. Accordingly, the meaning of responsibility dedicated to each single employee must be underlined. In parallel, details about correct and incorrect behavior connected to cause and control should be outlined to generate appropriate awareness.

6.3.5 Premises and Equipment

The guidelines demand a regular cleaning interval of premises and equipment. Moreover, suitable procedures must be implemented to prove contamination evidences. Their incidence has to be documented and relevant measures need to be defined. The cleaning, disinfecting, or sterilizing operations should follow written instructions. The precision of these instructions as well as the cleaning personnel and the installation of premises or equipment critically determine the cleaning outcome. Therefore, hygiene planning must cover and define the following topics:

- Subject of the cleaning (surface, equipment, floor, etc.)
- Frequency of cleaning (e.g., monthly, weekly, daily, or more often if required)
- Cleaning agent, its usage (e.g., exposure time), and preparation
- Type of cleaning equipment (sheet, mop, etc.)
- Cleaning method (wiping, disinfection, sterilization, etc.)
- Time-point of cleaning (e.g., after maintenance, repair, batch, or product change)
- Safety instructions related to the cleaning procedure
- Documentation of the cleaning activity

The manufacturing area (at least surfaces with contact to the product and floors) of CBM and the respective equipment should be the subject of cleaning after each usage. In grade A and B, only sterile cleaning agents should be used, and a suitable exposure time must be ensured. Due to the nature of CBM and the aseptic processing needs, only agents with suitable disinfecting activity must be used. Surfaces and equipment might be wiped with tinctured non-fraying sheets, while mops with sterile wipers are appropriate for the cleaning of the floor. Cleaning activities might be documented at the end of the manufacturing record.

Depending on the facility design and the type of production (campaign or single products), the manufacturer has to define measures to prevent cross-contamination, which has to be considered for cleaning activities, likewise. Nevertheless, a cleaning interval for the professional disinfection of the complete facility including walls, all surfaces, and ceilings should be defined as well. It is advisable to use agents with a different mode of action to avoid resistances and personnel that are specifically trained for this purpose.

It should be demonstrated and validated that the cleaning procedure is adequate for the manufacture of the specific CBM product. Furthermore, appropriate cleaning must be monitored and documented. This monitoring should be carried out according to predefined acceptance criteria, warn, and alarm limits. The compliance of acceptance criteria should be trended, while corresponding measures must be defined if warn or alarm limits are exceeded. Microbiological monitoring activities must control all surfaces of the clean room area, the personnel, and the air. Commonly swabs, suitable air sampler, settle and contact plates are employed in this context. The sampling volume, length, and limits according to the cleanliness class are predefined in the guidelines. Nevertheless, a detailed sampling plan has to be specified. Thereby, the manufacturer has to consider the results obtained during clean room qualification to determine relevant sampling points.

The hygiene concept for the clean room surrounding area has obviously less stringent requirements. Nonetheless, the cleanliness level is similarly of importance, specifically for the QC rooms, the stores, and as flagship for cooperation partners or inspectors. QC rooms should be adequately cleaned to avoid a negative impact on product controls and analytical procedures that are employed to determine product release relevant characteristics. This accounts, of course, for the specific equipment in parallel. The cleaning of storage rooms including inventory like freezers is essential since it significantly influences the materials that are further used for the aseptic manufacturing process of CBM. Finally, the hygiene concept should also schedule the cleaning operations for the remaining supportive area.

6.3.6   Materials

Materials should not only be checked for damages or other quality impacts, it should also be cleaned or freed from outer packaging (if required). Deliveries ought to be inspected regarding adequate cleanliness before being freighted to the stores. The removal of dirt helps to sustain the quality of these goods, ensures less operational pollution of storing areas, and reduces the risk of (cross-) contaminations during the manufacturing process later on. Particularly, materials without removable clean-room-specific outer packaging should be clean to avoid the carry-through of dirt particles toward the rooms with higher cleanliness grades.

6.3.7   Pest Control

Respective measures must be defined to effectively protect the clean room facility against invading insects, for example. Accordingly, insect traps like adhesive tapes behind doors or traps based on ultraviolet light can be employed. These measures are particularly useful during respective seasons or in countries with climatically related increased relevant burden. However, pest-related incidents must be documented and monitored. These results can provide the basis for preventive procedures.

7. Manufacturing

The content of this section is based on respective regulations and guidelines that provide further specific and profound information [17,18,26,27,34–40].

The manufacturing of goods can be diversely described and defined. Many definitions are listing actions like extracting, producing, preparing, processing, filling, packaging, labeling, and releasing to further specify this activity. However, the common denominator of all these definitions is the aim to describe the process of manufacture. Correspondingly, the production of CBM should be understood as a manufacturing process with an emphasis on the word process. Generally, in Europe, the manufacturing authorization, which is approved by the respective authorities, is a prerequisite if CBM products are envisaged for clinical studies or subsequent market authorization. Therefore, all single operations of the complete manufacturing process must be carried out and controlled by competent personnel in accordance with clearly defined procedures and the principles of GMP. Solely on this basis, CBM products are considered to be of appropriate quality and to comply with the relevant manufacturing or market authorization demands.

Further deliberations aim to follow and concentrate on a logical sequence of all manufacturing operations of CBM and are tightly coupled to the material flow.

7.1   Materials

There should be written specifications for all materials to ensure, for example, their proper identification, storage, and approval. These specifications might provide information about the following material characteristics:

- Intended purpose, manufacturer, supplier, container size, and number
- Storage instructions, lot number, date of expiration, and quality characteristics

Moreover, material specifications should define the corresponding inspection characteristics (e.g., appearance, purity, sterility, or endotoxin), inspection methods, and acceptance criteria. Results and confirmations of acceptance criteria are subjected to QC activities. They can be done by the manufacturer or in external contract labs. Under certain circumstances, these information might be obtained via certificate verification. The structure of the specification document may also enable the recording and approval of these issues in parallel. Nonetheless, materials must be approved by the QC department before they are released for the manufacturing process.

Materials that are intended for the manufacturing of GMP-compliant products can occasionally be purchased with or without "GMP-grade" classification. Nevertheless, the manufacturer has to ensure that the material fulfills all GMP- and product quality-relevant characteristics irrespective of whether it is certified or not. This implies, for example, that material and ingredients are traceable and that they do not pose any risk to the patient or product quality.

7.2   Goods Receipt and Storage

The receipt, identification, quarantine, storage, handling, sampling, testing, approval, or rejection of materials must be outlined in written instructions. All materials are subjected to incoming goods and documentation control. Received containers must be cleaned (if necessary) and labeled accordingly. Damages, impairments, or other problems that can influence the product quality must be investigated, documented, and announced to the QC department. Incoming goods must be held under quarantine until specifications are checked and the material is identified as appropriate. Otherwise, they should not be mixed with existing approved materials to avoid cross-contamination and mix-up. The separation can be realized via different storage areas or other administrative measures.

Purchased intermediate products should be handled just like starting materials if they are intended to be used for the manufacturing process. However, all materials must be stored under adequate, organized, defined, controlled, and documented conditions to enable batch separation irrespective of whether they are under quarantine or approved for manufacturing.

7.3 Manufacturing Preparation

Not all materials can be directly applied to the manufacturing process. Some of them like highly concentrated solutions or agents may have to be diluted, for example. Such dilutions are often associated with the preparation of stock solutions and subsequent aliquot measures. Nonetheless, these aliquots have to comply with GMP regulations as they are intended for manufacturing activities. Consequently, they may have indirect or direct product contact with all its importance regarding potential quality risks. This means, for example, that the sterility of stored supplements that are not directly used but are intended to be employed during cell culture has to be demonstrated. The stability and storage conditions have to be investigated, documented, and defined in written specifications. Another important issue concerns the labeling of such aliquots. The label must be clearly readable, permanent, and contain the following information:

- Name of the base material, concentration, and diluent (if applicable)
- Lot number of all base materials
- Date of manufacturing and expiration
- Initials or signature of the manufacturing person
- Storage temperature

Starting materials, agents, or solutions for repeated usage additionally have to be labeled with the following items:

- Batch number and in-use stability
- Date of beginning and initials of the employee

7.4 Clean Room Classes

The manufacturing of CBM is commonly based on aseptic procedures to ensure the sterility of the finished product. Nonetheless, not all processing steps might require aseptic conditions. Accordingly, the different grades of clean rooms (see Table 1) and examples for their intended use can be outlined as follows:

- Grade A:
  The highest level of cleanliness is named grade A according to EMA regulations. This class is reached in laminar flow cabinets. Their laminar flow profile has to be validated and should

range between 0.36 and 0.54 m/s. Grade A is supposed to be used for aseptic preparations, filling, or handling steps on open systems like vials or culture dishes. It should be continuously monitored for particles during operation.

- Grade B:
  Cleanliness grade B is following grade A and is the next lower clean room class. Whenever aseptic processing is scheduled, it must be done in grade A system. In case this is an open system (e.g., like a laminar flow cabinet) it must be embedded in a grade B environment. Continuous particle monitoring in operation is suggested.
- Grade C:
  Clean room areas of grade C are intended for less critical manufacturing steps of sterile products. Less critical steps include the preparation of (intermediate) products that can be sterilized in the final container or solutions that can be filtered. Monitoring of particles during operation is not binding.
- Grade D:
  Cleanliness class D is the lowest defined clean room grade. It is also intended for less critical activities, for example, the handling of components after washing. Particle monitoring in operation is not required.
- Isolator:
  The isolator technology aims to minimize the risk of environmental contamination. An isolator is an encapsulated area including locking system for the entry and exit of materials. The transfer of materials bears the highest risk for contamination. Therefore, air quality and integrity have to be assured by validation and/or leak testing. Isolators must at least be surrounded by cleanliness class D.

The EMA and FDA classifications of clean room grades have different requirements. In Europe, the EMA defines four classes (A–D) and two types of particle limits (at rest and in-operation) for two particles sizes ($\geq 0.5\,\mu m$ and $\geq 5\,\mu m$). The FDA specifies only three different cleanliness grades (100, 10.000, and 100.000) and a single particle limit (in-operation) for a single particle size ($\geq 5\,\mu m$). Class 100.000 and 10.000 equal grades D and C, while class 100 correlates to grades B and A. Particle limits at rest, as defined by the EMA, should be reached after approximately 20 min. This provides some guidance regarding the required air exchange rate in clean room areas.

7.5 Manufacturing

The manufacturing of CBM is critically relying on aseptic processing due to the fact that these products cannot be sterile filtered or sterilized without detrimental effect on the product quality. Production processes of CBM often necessitate multiple different manipulation steps on "open" systems like cell culture dishes, for example. Moreover, frequently, the source of starting material is limited due to the human origin. Consequently, an expansion of this cell material is required to obtain therapeutically relevant cell numbers. This processing usually takes several days or weeks, during which the cell culture needs to be supplemented with fresh medium, for example. Therefore, reliable accurate working methods of all personnel as well as cleanliness of premises and equipment are basic requirements of all manufacturing activities. These requirements are further stressed due to the fact that antibiotics are commonly not allowed during cell cultivation of CBM. Otherwise, traces of antibiotics will be part of the finished product with potentially critical impact. Moreover, they might suppress the outgrowth and expansion of hazardous organisms during cultivation. Hence, the amount of contaminating entities might be too low and impede their detection due to a low probability of detection. But, certainly microbes will rapidly expand if antibiotics are waning, which is the case after infusion to patients.

All materials that are required for a single process step have to be gathered and checked for their expiration date before being transferred to the clean room area. The route of entrance should be different than the personnel to avoid contamination in locking rooms where garments are changed. Instead, the facility should be equipped with separate material locks of adequate size. It has to be kept in mind that equipment of large size, like microscopes, ladders for maintenance work, or defective centrifuges, needs to be entered or exited too. However, all things must be carefully cleaned before they are placed in the material lock. Accordingly, wrappings must be removed if necessary, and the exposure time of disinfecting agents must be guaranteed. In case sterilized materials have multiple secondary packaging, only the outer one should be removed. These wrappings are intended to be removed stepwise while material is transferred to the next cleanliness class. Generally, these locks have to fulfill all previously mentioned GMP requirements. Briefly, walls, ceilings, and floors must be easily cleanable to ensure a low particulate burden. Therefore, pipelines and lightning systems must be encapsulated. Moreover, the air exchange of these rooms should reach the envisaged cleanliness class in a suitable timeframe. Depending on the size of the lock, it might be useful to indicate the borderline between lower and higher cleanliness grades.

Nevertheless, the personnel have to enter the clean room area via different locking systems. These employees must be healthy and trained for the scheduled activities. Besides manufacturing operations, there should be a special focus on garments (changes), cleaning, behaviors, and prevention of contamination. If the employee has changed their clothes and entered the respective clean room area, the material lock can be opened after air exchange is completed. Afterward, all materials can be transferred to the next lock. Still, materials have to be disinfected again if they are not further wrapped. The subsequent proceeding corresponds to the previous actions. After reaching clean room class B, materials can be removed from the locking system and employed to the manufacturing process. The clean bench should be disinfected before its use. In this context, exposure time of disinfecting agents and the run-in period of the bench have to be considered. All other equipment that is required for the process step should be checked for its functionality to ensure consecutive work flows. Thereby, the exposure time of the cell product to nonideal conditions can be minimized. The lot identification of all materials that are used for processing as well as all manufacturing activities must be documented according to written instructions. Furthermore, materials that are scheduled for aseptic work need to be disinfected before they are transferred to the clean bench.

GMP compliance of aseptic processing requires environmental monitoring. Accordingly, the number of small ($\leq$0.5 µm) and large ($\geq$5 µm) particles as well as an air germ collection must be done (see Table 1). Additionally, the monitoring should include settle plates for A and B environments and a glove print after work is finished. The identification and assignment of all monitoring results to the respective process step and operator must be assured. Aseptic processing has to avoid unnecessary changes between grades A and B. This has to be considered for the preparation of materials for consecutive steps if no second person for assistance is available. Otherwise, the employee has to carefully disinfect the gloves before re-entering grade A. However, this procedure is not appreciated by authorities since it can affect the product quality due to the carry-over of particles.

Once aseptic processing is finished, the cell product should be transferred back to the incubator as soon as possible to provide optimal growth conditions. The incubator should be dedicated to one product batch during its manufacture to avoid cross-contamination and the possibility of mix-ups. Besides, equipment that will not be used during the production process (e.g., redundant equipment like a second incubator) shall be indicated for the same reason. After the intended process step is completed, all monitoring plates have to be sealed to avoid their contamination due to cleaning activities. Prior to cleaning the waste (e.g., used pipettes, wrappings, etc.) is collected and transferred to the material lock. Afterward, equipment that was employed during processing is disinfected. Moreover, the clean room itself and all product contact surfaces have to be disinfected likewise. Additionally, at least the floor of all previously accessed areas shall also be disinfected. Preventive measures regarding cross-contamination and/or mop-exchange have to be defined depending on the size of the facility and the envisaged production processes.

7.6 In-Process Control

The basic idea of GMP compliance includes the principle that it is not sufficient to confirm quality just by testing. Instead, it has to be accomplished in a step-by-step process. Therefore, in-process control (IPC) or process monitoring is required to demonstrate this quality building process as well as process control.

CBM products are very distinct compared to chemically synthesizable drugs. The main difference is certainly the manufacturing process. Biochemical processes are precisely defined, and their management can rely on key indicators. Even the starting material has a consistent standardized quality that makes these processes largely predictable and enables a process management in strictly defined limits. The opposite is true for CBM. Already, the starting material is heterogeneous, and its composition is as unique as each individual from whom it is derived. Consequently, each batch of a CBM product will have its own characteristic, and the level of standardization will possibly never be comparable to common drugs due to the biological variance. Nevertheless, safety and efficacy issues of CBM are not of minor importance compared to any other therapeutic intervention. Currently, the manufacturing of CBM is often a lengthy aseptic process, and finished products are applied even if microbiological tests are not completed or their results are not yet available.

Therefore, critical process steps are subjected to validation, and IPC needs to be defined. IPC should cover all critical steps (if possible), unless the product quality is negatively affected. Nonetheless, it is advisable to monitor the process as closely as possible to demonstrate process control. Thereby, IPC can be extensively used as long as the process is not hampered. This means that certain IPC should only be scheduled if enough material is available and the required product yield can be guaranteed. It also implies that IPC methods should be validated to ensure data integrity, which is significantly important if they are used for process management purposes. The IPC sampling plan should be approved by the QC department. It defines how the sampling is approached with respect to the process step, sample size, and/or time point. Furthermore, potential retention sampling, acceptance, and action limits should be further specified. Accordingly, exceeded limits might require further tests, quarantine storage, or even a process stop, for example. However, retesting, for example, is only possible if retention samples of suitable size and quality are available. This necessitates an adequate planning with regard to storage conditions and sample size.

Nevertheless, IPC or process-monitoring activities are unique to each production process and shall be based on an RA. However, potential IPC candidates are listed in the following:

- Sterility test (e.g., after critical or extensive handling on open systems)
- Cell count and cell viability
- Cell doublings or telomere length
- Metabolic activity
- Phenotype or product composition
- Karyotyping
- ATP consumption

Some methods and their intention are further described in the next section.

7.7   Filling, Finishing, and Labeling

The filling and finishing of CBM products is a rather difficult process due to the difficulty in handling of the product containers. Commonly containers like freezing tubes or bags, blood bags, syringes, and vials made of plastic or glass are employed. These system have, for example, to be tested for the absence of pyrogens like endotoxins, abrasive particles, and sterility. Bag systems are intended to be filled via tubing connections, which requires sterile tube welding devices or Luer-lock connectors, for example. Consequently, these aseptic manufacturing steps need to be validated according to the GMP guidelines. Thereby, the maximum number of filled containers as well as worst case issues should be considered. Specifically, less automated filling steps by hand necessitate well-trained personnel with valid aseptic processing skills. The validation of these skills must be based on recurrent media fills.

In case vials are used that need to be crimped, the container closure system is not considered complete until crimping on the stoppered vial is done. Therefore, crimping should be realized as soon as possible after stopper insertion. This can be done inside or outside the clean room area, but vials must always be protected by grade A conditions (e.g., by laminar flow techniques). Usually, the batch size of CBM products is small, and the complete filling and finishing process will be done by hand. Nonetheless, container integrity should be carefully checked.

Finished products must be kept under quarantine until all release tests showed specification conformity and have been approved for release.

The labeling of CBM products is of upmost importance since products, for example, are often intended for autologous approaches and might otherwise cause fatal reactions in the patient. For that reason, filling and sealing should be followed as quickly as possible by labeling to avoid mix-ups. In the case when automated systems like a label printer are employed, their functionality must be assured and documented. Likewise, the number of approved and released labels should be recorded and subjected to balancing after labeling. Supernumerary labels must be destroyed. If labeling is a manual process, it must be checked by a second person. The quality of labels has to ensure their integrity for the intended storing conditions and shelf life. Storage under cryogenic conditions has to be considered in this respect. The label itself has to provide the following information:

- Name, address, and telephone number of the manufacturer
- Name of the product and batch number
- Route of administration and storage conditions
- Cell number or dosage and volume
- Date of filling and expiry
- Content and ingredients
- Warnings like "Keep out of reach for children"

Further (national) information may be needed, especially if the product will be applied during clinical studies. Accordingly, the space on the primary packing might become a limiting factor since the label has to ensure good legibility. Therefore, some of this information can also be provided on a leaflet, which then has to be issued on the label. All information should use the language of the federal state of application.

7.8 Packaging

The packaging process of CBM products should be primarily understood as the wrapping in outer or secondary packaging systems. Nonetheless, the purchase, handling, and control of all packaging materials are similarly important as the management of starting materials. Particularly, printed materials or labels should be access restricted and stored under secured conditions. Outdated or no longer required materials should be destroyed and their disposal the subject of documentation. Currently, CBM are typically not routinely applied. Therefore, the packaging process is frequently not standardized to a professional industry-like level. But, even if secondary packaging is rather simple, it should be of adequate material and support the quality of the finished product (e.g., with respect to the conditions during storage or delivery).

7.9 Contamination Prevention

Contaminations are a major thread for the manufacturing of CBM since this therapeutic option critically relies on sterile finished products. In this respect, contaminations comprise impurities that are detectable by microbiological tests but also cross-contaminations with other product batches, for example. Cross-contaminations can usually not be detected with microbiological methods and are potentially hazardous in case of an envisaged autologous application. Moreover, contaminated CBM products are extremely dangerous because they are routinely injected into the patient's bloodstream.

In general, potential contamination risks arise from the personnel, starting materials, air, premises, and equipment, but also from the design of the manufacturing process itself (e.g., regarding the use of closed instead of open systems). The prevention of contamination is an essential part in the stepwise generation of consistent product quality and GMP compliance. Therefore, and with respect to the hazardous potential of contaminated CBM products, some basic requirements should be summarized again. Premises and equipment should enable an easy cleaning procedure. They must be cleaned on a regular basis and should be dedicated during the manufacturing process. Their surfaces should be smooth, not particle emitting, and their arrangement should enable a logical process flow. The personnel have to be trained particularly with respect to hygiene aspects, behavior, and correct clothing. All required materials should be precisely specified and kept under good storing conditions. Supply and extract air must be filtered, and windows shall be closed to avoid the entry of insects. Moreover, an adequate airlock system and pressure cascades in the clean room area shall be installed. Finally, it should be underlined that the document system (regarding extent, style, etc.) is able to efficiently support the prevention of contamination risks.

7.10   Deviations

Deviations during the manufacture of CBM are often related to the complex, lengthy, and less standardized production process. They do not necessarily negatively impact the quality of the finished product. Nevertheless, their handling is of importance. Deviations must be reported, assessed, and corresponding measures need to be defined. The deviation description, investigation results, and improving or preventing measures should all be issued in a written format and the subject of a documented deviation report. This information is suggested to be incorporated in the annual product review for further trend analysis.

Deviation reports might arise from observations, defective equipment, expired materials, mistaken instructions, or documents. Their importance related to the product quality should be appraised. Seriously affected batches must be indicated, separately stored, and kept under quarantine until the deviation is evaluated and corresponding measures are defined.

7.11   General Issues

The manufacture of CBM can be divided in products that require a subsequent cultivation period and those that do not. Some products are based on cells that are already available in large quantities in the human body or whose production process includes a sufficient source of starting material (e.g., leukapheresis products). These cells just need to be isolated or enriched and can be directly applied or stored thereafter. In the case when starting material is limited, for example, due to the patients' health condition or if the desired cell population is too small to achieve therapeutic benefit, the manufacture relies on cell cultivation procedures. Obviously, cell culture must also be considered if cells are subjected to other modifications (e.g., genetic modification). Some general issues have to be clarified as soon as cell culture techniques are applied. These topics cover fundamental cell cultivation aspects. Nevertheless, their consideration might be differently important with respect to GMP compliance and its trends.

- GMP-grade materials
  Commonly materials like cell culture medium or serum are available with or without being labeled "GMP-grade" or "GMP-certified," for example. Aspects like purity, stability, or composition are often equal, and their suitability must be assured by the manufacturer irrespective if certificates are available or not. However, usually the accompanying documentation is very different and might enable reduced test scopes. Sometimes, these tests are even hardly possible for the manufacturer. This is particularly valid if ingredients of medium or the existence of certain pathogens in the serum need to be determined. Therefore, comprehensive and detailed concomitant documentation might be preferable.
- Serum versus serum-free cultivation
  The use of serum for the supplementation of cell culture medium is still indispensable. Although serum-free condition have been the subject of extensive research for decades and have been established for certain cell types and applications. Serum-free conditions are of advantage since replacements have a defined composition and batch-to-batch variances as well as the risk of pathogens (e.g., Transmissible Spongiform Encephalopathy [TSE]) are smaller. Thereby, the manufacturing process becomes more defined, better manageable, and the overall risk is decreased. Accordingly, authorities favor serum-free conditions. They are still not officially required, but every manufacturer should consider this topic in future times. Consequently, manufacturing processes that are under development or just envisaged should be carefully checked for a serum-free realization prior to their GMP-compliant translation.
- Closed systems versus open systems
  Traditional cell culture dishes are open systems that are susceptible to contamination. Nonetheless, they are widely used in research. Closed systems like bioreactors exist, but they are less frequently employed. Process translation to bioreactors is often difficult since multiple parameters (e.g., culture surface, gas exchange, etc.) are changed and need to be carefully adapted or modified. Sometimes, retrospective process translation fails due to unexpected cellular changes regarding functionality, phenotype, potency, or metabolism, for example. This underlines that the manufacturer should approach the potential use of closed cultivation systems from the beginning. Many systems are already available particularly for cells growing in suspension or that are available in sufficient quantities to enable a successful inoculation of these systems. However, the trend clearly moves toward closed systems. Regulatory and safety reasons are undisputed, but technical advantages are likewise significant. The adaption to a closed system is usually accompanied by a simplified process control and management that is less time and labor intensive.

8. Quality Control

The content of this section is based on respective regulations and guidelines that provide further specific and profound information [8,34–36,41–45].

Each manufacturer must have a QC department that has to ensure all relevant sampling and testing operations are carried out. Correspondingly, specifications, documentation, and release procedures must be prepared. The QC unit is not only concerned with laboratory operations. Instead, it must be involved in all product quality-relevant decisions. Accordingly, manufactured CBM products are only released if the QC approves their satisfactory quality. Furthermore, GMP guidelines stipulate a strict separation and independency from the manufacturing department to assure the autonomous operation of the QC department.

The main features of sampling, quality planning (e.g., regarding specifications), and testing, which are in the responsibility of the QC unit, can be exemplified as follows:

- All measures and tools of an effective and reliable control of verification requiring products must be assured. These measures include the approval of suppliers, definition of verification requiring materials, as well as the control, release, or rejection of starting materials, packaging materials, intermediate, and finished products. The latter is done according to predefined written specifications that must be satisfied. Moreover, it comprises the labeling and control of all starting materials, packaging materials, intermediate, and finished products according to their status (approved, released, or rejected). Sampling and the handling of retention samples is covered, too.
- The validation of all quality-relevant test methods, as well as control and approval of IPC methods employed during manufacturing, including respective planning and documentation, has to be realized.
- Stability (re-) testing of starting materials, intermediate, and finished products must be defined and regular product quality reviews scheduled.
- Quality deviations, returns, contracting, and audits should be processed in cooperation.
- All procedures must be approached and documented according to predefined written instructions.

Generally the QC personnel must be authorized to access all relevant areas to perform the sampling of materials as well as intermediate and finished products. The following parts refer to some GMP and QC relevant issues which were outlined previously. Nonetheless, important aspects are revived and supplementary QC-related facets are added.

8.1 General Policies

Premises and equipment of QC labs should be in accordance with all general requirements (e.g., cleanliness) that were described for QC areas in the corresponding chapter "Premises and Equipment." The personnel as well as premises and equipment should both be suitable for the intended purpose. Accordingly, QC employees must be trained (e.g., regarding specification limits, test methods, etc.), and the department capacities should consider the scale of duties. Contract laboratories are accepted for certain controls like endotoxin or sterility test, for example, but this must be addressed in QC records and specified contracts. Equipment that is employed for QC activities must be qualified, maintained, and calibrated. Calibration can be defined as comparison of a measured value and true value under defined conditions. Deviations from the true value can be corrected by adjustment.

8.2 Documentation

All QC-relevant activities including their documentation should be done in accordance with written instructions as outlined in the chapter "Documentation." GMP compliance requires the documentation of many different things and activities. Essential parts of these documents must be made available for the QC unit, including the following:

- Specifications and sampling procedures
- Test procedures and test records (including analytical worksheets and/or lab books)
- Analytical records and/or certificates
- Instructions for the calibration and maintenance of equipment including respective records
- Records concerning validation of test methods and environmental monitoring (if required)

QC documents that are related to a batch record must be retained for 30 years after expiry or clinical use for ATMPs like CBM products. Data like analytical test results or environmental monitoring should be kept in a manner that enables trend analysis. Likewise, other original data originating from lab books or records should be retained and easily available.

8.3 Sampling

Sampling is an important operation since it provides information about the current manufacturing process status, for example. The process management is often based on these data. Consequently, it is of significant importance that corresponding samples are representative. Otherwise, they may lead to false conclusions and cause a negative impact on the production process. Hence, correct sampling is an indispensable part of the QC, QA, and process management. Obviously, the personnel must be well and continuously trained to gain an adequate level of experience for this meaningful operation.

The sampling commonly concerns, for example, starting and packaging materials, IPC, intermediate and finished products, hygiene, monitoring, or stability issues. Additionally, different methods like liquid, solid, single, or mixed sampling might require different procedures like sample homogenization or precautions regarding light or temperature sensitivities of the samples. Accordingly, diverse sampling containers are available. They should be clearly specified with respect to size, form, material, sterility, pyrogenic, or particle content, for example, to avoid falsification of the results. Moreover, the product quality (e.g., regarding sterility) and the labeling of tested container must be ensured. In this context, specific garments, equipment, and auxiliaries, for example, shall be defined likewise. Therefore, sampling should be carried out according to approved written instructions, which should cover the following:

- Sampling method and corresponding equipment
- Sampling volume and instructions for further sub-aliquots

- Type and condition of the sampling container
- Labeling of samples and processed materials or container
- Precaution instructions (e.g., regarding sterility and cross-contamination)
- Storage conditions of samples
- Cleaning and storage of sample equipment

Sampling might be extended to control critical process steps like the process initiation or the filling. All sample containers must be labeled with information about content, batch number, date, and entity from which it has been taken.

8.4 Sampling of Process Materials

The sampling of process materials like medium or supplements (e.g., serum) should ensure parameters like identity and quality. Normally, this can only be achieved if samples from all containers are taken and all of them are analyzed. Accordingly, it might be suitable to prepare a representative sample to reduce the work load. The preparation must be specified in a sampling plan with respect to the number of individual samples that have to be drawn. Such an approach should be justified and may be subjected to risk analysis and statistics. The extent of sampling might be further reduced if the supplier has been qualified in advance.

8.4.1 Sampling of Packaging Materials

The number of samples for packaging materials should be statistically determined and the subject of a sampling plan. This plan should consider the required quality and quantity of the received material. Additionally, the intended use (e.g., primary or secondary wrapping) and knowledge gained about the manufacturer during audits should be reflected.

8.4.2 Reference and Retention Sampling

There are two types of samples that are intended to be retained. They should be kept to provide a sample for analytical procedures and a specimen of the finished product.

The reference sample is drawn from a batch of material or product and intended to be used for analysis during the shelf life of the concerned batch, if this is required. The retention sample is defined as a sample of a packaged unit from a finished product batch that is stored for identification purposes.

In case of finished products, both sample types will frequently be identical. They can be regarded as part of the batch record and are assessable if the product quality is concerned for certain reasons. Consequently, their traceability for further review by competent authorities must be assured.

The sample size should be sufficient to allow the complete spectrum of release testing. It might differ depending on the batch size and intended application. The need for re-testing, for example, might be higher if a large batch size is envisaged for an allogeneic approach compared to a small batch size that is scheduled for an autologous treatment. Nevertheless, stability issues, analytical problems, or complaints should be considered. Therefore, the sample size has to be defined and must at least satisfy the regulatory requirements. Moreover, the sample management regarding responsibilities and documentation must be precisely specified.

The storage condition must ensure an adequate quality of reference and retention samples from finished product batches for at least one year after the expiry date. In case CBM products are employed in clinical studies and considered as investigational medicinal products, they have to be kept for at least two years after completion or formal discontinuation of the last clinical trial in which the batch was applied. The storage location of these samples, which usually has to be at the manufacturer side, must be defined and agreed upon between sponsor and manufacturer under this circumstance. Starting materials must be kept for at least two years after the release of the final product, unless national laws require a longer period or the expiration date is shorter. Packaging materials shall be retained according to the duration of the finished product.

8.5 Testing

Analytical procedures that are employed for testing must be validated, and the respective equipment has to be qualified accordingly. Commonly, analytical methods are considered validated if they are described in a pharmacopeia. Cell culture medium and supplements that are used for the manufacture of CBM products are usually not included. Therefore, the manufacturer has to ensure that the validation covers the product- or production-specific matrix.

The testing approach should be approved by the QC department and authorities. It should be carried out accordingly, and the respective record should cover at least the following:

- Name of the material or product and dosage form (if applicable)
- Batch number, manufacturer, and/or supplier
- References to relevant specifications and test procedures
- Test results (including calculations, observations, and if available, reference to certificates of analysis)
- Dates of testing and initials of the performing personnel
- Initials of test or calculation verifying personnel (if appropriate)
- Precise statement of the status (e.g., release or rejection) including date and signature of the responsible person

The quality of laboratory reagents, volumetric instruments or solutions, reference standards, and culture media must be assured. Their preparation and usage must be done according to written instructions. Laboratory reagents that are envisaged for a longer usage must be signed by the preparing person with date and signature. Expiration date and storage conditions must be indicated on the corresponding label. The date of receipt should be indicated on all substances that are employed for testing if this is required.

Reference standards can be used to compare qualitative and quantitative methods, for example. They are, nonetheless, often not available for CBM products due to their specific characteristics. However, if they are available, they have to be acquired via official and authorized sources (e.g., like chemical dealers or pharmaceutical manufacturer). Accordingly, substantial accompanying documentation similar to that of the starting materials is needed. The labeling of such substances is equally important. All issues account for internal and in-house manufactured standards as well. Finally, it has to be ensured that all required analytical equipment or materials are still available for at least one year after the shelf life of the manufactured batch.

8.6   Product Stability

The stability of CBM products should be controlled according to an appropriate continued approach that enables the identification of questions and problems related to stability, dosage form, or packaging. Accordingly, this program aims to verify that the product complies with its specifications under the defined storage conditions throughout the complete shelf life. Intermediate products that are stored and used for longer periods shall be tested likewise. The ongoing stability program and its results should be the subject of a written plan and report. Employed equipment must be qualified. The schedule of an ongoing stability program should cover the shelf life endpoint and include the following information:

- Number of batches, reference to the test method, and acceptance criteria
- Relevant physical, chemical, microbiological, and biological test approaches
- Envisaged test interval and other specific product relevant parameters
- Description of the container closure system and storage conditions

The number of investigated batches and the frequency of testing should provide an adequate data basis to enable trend analysis. At least one batch per year should be subjected to stability testing. Results must be made available to key personnel and the QP. In case results are out of specification or significant atypical trends are recognized, corresponding investigations must be initiated. All data including preliminary conclusions must be summarized in a written report. This document shall be maintained and periodically reviewed.

8.7   General Quality Control Methods

Methods employed for QC must be validated, and the respective equipment or instruments should be qualified. The approach of these activities including the required documentation has already been outlined. There are many general QC methods that are applicable to all CBM products and even more particular ones that are product specific. Nonetheless, each manufacturer has to define the product-specific QC techniques that are required to demonstrate a consistent manufacturing process, sustained product quality, and minimized patient risks. This might be achieved after consultation and/or agreement of the competent regulatory bodies.

8.7.1   Sterility Testing

The manufacture of CBM critically relies on aseptic processing to generate sterile finished products. Aseptic processing is validated with the help of media fills that use a specific growth-promoting medium. Sterility of media fills and finished products can be assessed in different ways. Corresponding techniques are delineated in the European Pharmacopeia Chapter 2.6.1 and 2.6.27 [46,47]. The use of blood culture bottles is a very suitable detection system, since most of the CBM product will be injected to the patient's bloodstream. It has a higher sensitivity and broader detection range compared to other sterility tests, like the membrane filtration or direct inoculation technique. Moreover, it decreases the detection time from two weeks to one if an automated system is used. This system is based on an indicator change that is caused by a growing microorganism. Microorganisms are able to grow under aerobic or anaerobic conditions. Consequently, each test is composed of two bottles, one for aerobic and one for anaerobic detection, which must be continuously monitored during cultivation. The indicator change is initiated by CO_2 production or oxygen consumption. It either leads to a color change of a pH indicator (BacT/Alert® system) or a quenching of a fluorescent dye (Bactec® system). Nonetheless, it has to be assured that the system is working with the specific matrix of the corresponding CBM product. Every testing should include positive and negative controls to demonstrate reliable and valid results.

8.7.2   Mycoplasma Testing

Mycoplasmas are very tiny bacteria without cell walls and lack different metabolic pathways due to their small genome. Accordingly, they are difficult to detect by microscope, insensitive against common antibiotics, and able to pass the common sterile filters. Mycoplasma is a human pathogen causing, for example, diseases of the bronchial and urogenital tract. Contamination of cell cultures can cause metabolic and chromosomal aberrations of the cells. Infections are mainly caused by cell culture supplements like animal serum or personnel, particularly in open culture systems.

Detection techniques for mycoplasma are outlined in the European Pharmacopeia (Chapter 2.6.7) and can be based on different cell culture methods (e.g., with or without indicator agents) [48]. These methods often require an extensive cultivation period of up to three weeks, which might be inconvenient if the CBM product is urgently needed. Polymerase chain reaction (PCR)-based techniques represent a much faster way for the identification of mycoplasma. The nucleic acid amplification can be quantitatively analyzed if a real-time PCR method is employed. Nonetheless, these techniques have to be validated to ensure that the method is not negatively influenced by cell culture matrix components.

8.7.3   Endotoxin Testing

Endotoxins are a major thread of parenterals due to the causing of toxic reactions in the patient. The human body tolerates only very limited amounts of these substances. Endotoxins belong to the chemical class of lipopolysaccharides (LPS), are very resistant to heat, and will survive sterilization procedures. LPS originate from the cell wall of gram-negative bacteria and cannot be removed from finished products. Consequently, endotoxin-free products can only be ensured if all employed materials are endotoxin-free and appropriate manufacturing techniques are applied.

The detection of endotoxin is issued in the European Pharmacopeia (Chapter 2.6.14) [49]. Endotoxin contamination is commonly proven with the limulus amebocyte lysate (LAL) test. This method is based on the finding that lysate from the horseshoe crab (*Limulus polyphemus*) results in jellification, precipitation, or turbidity if endotoxin is present. Nevertheless, this reaction can also be interfered by components of the CBM product matrix that has to be analyzed. Therefore, it must be validated for the specific cell culture medium.

8.7.4 Transmissible Spongiform Encephalopathy Testing

Transmissible spongiform encephalopathy (TSE) causes a degeneration of the central nervous system via infectious particles that are known as prions. Prions are very resistant and cannot be eliminated by sterile filtration or other sterilization methods like autoclaving or radiation [50]. Creutzfeldt-Jacob disease, for example, is a known human type of TSE. In animals, TSE is known as bovine spongiform encephalopathy (BSE) in cattle or Scrapie in sheep and goats. However, under certain circumstances, species barriers can be crossed [51].

Animal-derived materials, like bovine serum as a supplement of cell culture medium, are frequently employed during cell cultivation. Nonetheless, reliable testing methods providing valid results are not available. Therefore, it is of significant importance that only materials from animals of traceable TSE-free herds including respective documentation are used for the manufacturing of CBM. Freedom of TSE is commonly considered in relation to the geographical origin of these animals like bovine serum from Australia and certified by the European Directorate for the Quality of Medicines and HealthCare (EDQM).

8.7.5 Viral Safety Testing

The viral safety of CBM products is a complex field with respect to QC questions. This complexity is related to the diversity of existing viruses, corresponding detection techniques, and the source of materials (e.g., animal or human origin) that are used for the manufacturing of CBM. Accordingly, available detection techniques rely on different methods and comprise a large variety of viral pathogens with partially unique specificities. Nonetheless, the chosen detection approach must be adequate for the intended purpose. Therefore, each manufacturer is recommended to carry out an RA that considers the diverse factors that may impact the viral safety of the finished product and the risk to the patient, likewise. The risk assessment should, for example, cover factors like the origin of materials, donor history, infectivity, and pathogenicity with respect to the intended indication and recipient population. Additional issues have to be addressed if the CBM product is based on gene manipulations. This includes the manufacturing route of the vector as well as the vector product that will be used for the manufacture of the CBM.

Currently available detection methods are based on the infection of virus-specific target cells in culture and the subsequent analysis of these co-cultures. Further assays for the detection of viral pathogens employ immunofluorescence, enzyme-immuno, or immunochromatography techniques, for example. Probably the most suitable technique is based on the amplification of nucleic acid in polymerase chain reactions. However, it must be considered that not all viruses are detectable with each method.

8.7.6   Tumorigenic Testing

Commonly, the quantity of available starting material and isolation techniques determine whether the batch size allows the direct application of a CBM product. In case missing yield hinders direct therapeutic benefit, isolated cells need to be expanded in vitro. Cell expansion is considered a substantial manipulation of the cells. Prolonged expansion periods, respectively, and increased number of substantial manipulations is obviously associated with extensive cell divisions. Each cell division bears the risk of cellular transformation and, thus, tumor formation due to chromosomal instability. Consequently, chromosomal integrity should be determined if the respective cell type is known for such tumorigenic potential. Affected cell types are, for example, induced pluripotent or human embryonic stem cells.

Currently, chromosomal stability is most commonly determined via karyotype testing. Nonetheless, new techniques like Array-CGH are upcoming. Karyotyping is carried out by counting cells that have previously been arrested in the metaphase of the cell cycle.

8.7.7   Product Quality Characterization

The manufacturer of CBM products is obliged to show product safety to ensure consistent product quality and to demonstrate GMP compliance. Nevertheless, safety and quality issues also concern product characteristics that are related to the envisaged therapeutic effect of the CBM in vivo after application to the patient. Consequently, appropriate test methods (and animal models) must be available and validated. If this is not the case, these characterizing tools are the subject of internal research and development activities. Therefore, it is strongly recommended to consider adequate approaches already during preliminary or research work and before translation to GMP-compliant procedures is initiated. The following QC aspects need to be answered before GMP compliance and manufacturing or market authorization can be addressed. Although, the realization of these characterizations is very product specific and can only be estimated or defined by the manufacturer itself. The suitability of these tests should be based on scientific argumentations and agreed upon by competent authorities.

8.7.8   Viability Testing

The determination of viability is required since usually only viable cells will be able to exhibit therapeutic benefit in the patient. Moreover, dying or dead cells might realize factors that initiate immunological responses that can be detrimental to the envisaged treatment advantage. In parallel, cell viability is also an indicator for the quality of the manufacturing process. Nevertheless, the amount of viable cells will be batch dependent to some extent due to the biological variability of the starting material. The limits must be specified by the manufacturer.

There are several very different techniques available for the determination of viability. Trypan blue staining is one possibility to differentiate between life (colorless) and dead (blue) cells. This procedure can be done manually, but it is also part of different automated approaches. Other automated systems determine resistance changes in an applied electric field or different staining techniques to detect viable and dead cells.

8.7.9 Identity Testing

The identity of CBM is commonly determined by phenotypic analysis. A detailed phenotypic characterization is generally impossible without the staining of characteristic cell surface molecules. Many cell-specific surface structures are listed according to the cluster of differentiation (CD), like the CD3 molecule as a main characteristic of T lymphocytes. In case certain cell populations are not identifiable by just a single marker, they can be combined with others to allow a specific identification. Since these structures are typically very tiny, the staining approach often uses, for example, specific antibodies that are coupled to fluorescent dyes to detect and mark them. The analysis and quantification can be carried out manually or with the help of automated systems. Manual investigations are done by microscope. The microscope technique can be automated in the form of high-content screening machines, which are very suitable for adherent cell types, for example. Further automated techniques, like fluorescence activated cell sorting (FACS, often used as synonym for flow cytometry), are based on a fluidic system to quantify specific marker or marker combinations in cell suspensions.

Autologous CBM products must also be analyzed for unique or patient-specific characteristics. This can be done by genetic profiling or by human leukocyte antigen (HLA) typing if the cells (e.g., lymphocytes) express these individual molecules, for example.

8.7.10 Purity Testing

Purity testing of CBM products is closely associated with the identity of cells, and both can usually be approached in parallel. The applied methodologies are commonly identical. Testing for product purity does not necessarily indicate that a product or even the active part of a product is allowed to contain only one specific cell type. However, purity has to be demonstrated, and limits need to justified (e.g., based on animal data). These limits already imply the second association of purity testing. Also, potentially contaminating impurities that might negatively affect the quality of a CBM product or increase the patient's risk have to be determined. The extent of such analysis as well as corresponding acceptable limits should be the subject of RA and necessitate scientific justification. Limits will be product- and indication-specific and have to be defined. Moreover, the manufacturer should consider whether cell types that are not part of the active substance are automatically impurities in a sense of a negative impact on the treatment. Although no corresponding justification can be found in the guidelines, it should be worth discussing this issue with the authorities. Respectively, an example might be the presence of (few) natural killer cells in otherwise (antigen) specific T cell products.

Besides cellular impurities, CBM products might also contain noncellular impurities like cell proliferation promoting beads, for example. Obviously, the freedom of such impurities including appropriate validated test methods and depletion strategies need to be demonstrated too.

8.7.11   Potency Testing

The potency of a CBM product might be defined as its ability to exhibit biological or therapeutic relevant functions. Currently, potency testing is perhaps the most challenging task of all QC-relevant approaches. This can be related to the following facts:

- The mode of action for the respective product might be unknown.
- The mode of action is known, but appropriate methods are not available or need to be developed firstly.
- Methods are available, but validation is not possible due to susceptible methodologies.
- Methods are valid, but the biological variances are too high.

The latter also implies that the chosen assay might still not be adequate. Generally, potency can be assessed in vivo or in vitro. In vitro assays are preferable since readout, time intensity, and logistics are usually facilitated. However, potency might be associated with one or several of the following examples:

- Release of specific molecules like cytokines or cytotoxic substances
- Up-regulation of specific marker or marker combinations
- Certain regenerative potential
- Cell lysis or killing of specific targets
- Consumption of specific substances like adenosine tri-phosphate or cytokines
- Inhibition of cell proliferation

These exemplified characteristic functionalities can be approached with methods like enzyme-linked immune spot (ELISpot), flow cytometry, high-content screening, microscopy, or co-cultivation assays. Despite the diversity of these various methods, it must be considered that the final candidate has to allow its validation according to all relevant parameters that are outlined in the "Qualification and Validation" section.

8.7.12   Product Stability and Comparability Testing

The aforementioned defined test methods and their specified acceptances criteria will be employed for the demonstration of product stability and comparability. Stability data are required to determine the shelf life of a product with respect to the specified storage conditions of the finished product (e.g., cryopreservation). During the shelf life, all quality-relevant product characteristics as determined by the envisaged release tests have to fulfill the predefined specifications. The demonstration of product comparability is needed once changes related to the materials, production process, or storage conditions are made. If changes are implemented, the product is considered to be different unless comparability is shown. Comparability can be proven if all relevant release tests demonstrate the compliance with the product specifications.

8.7.13 Basic Quality Control Methods

Basic QC methods can be likewise engaged to monitor and manage the manufacturing process of CBM products. Macroscopic observation can provide important information about the metabolic activity (pH indicator change) or potential contamination (turbidity) of the cell culture, for example. Microscopic analysis, for example, gives insight regarding cell growth (confluence), morphology (shape), or overall status (granularity). Another basic, but regular and often used, QC approach employs the determination of the cell number. The cell number is an accepted process management parameter and usually is approached in combination with viability testing. Commonly, this method will be release relevant since most CBM products are applied in specified dosages. Certainly, if this is the case, the method must be validated as well, irrespective if hemocytometer or automated systems are used.

9. Inspections, Audits, Complaints, Recalls, and Returns

The content of this section is based on respective regulations and guidelines that provide further specific and profound information [9,52–54].

The manufacturer of CBM has to assess a bunch of external and internal aspects to demonstrate the competence of manufacturing CBM under GMP-compliant conditions. External aspects can be related to the supplier of process materials or contract laboratories, for example. Internal subjects might concern general topics like the QM system or specific issues like the training of certain methodologies. The required information can be gained in internal (self) or external inspections and audits. The segregation of inspections and audits is, to some extent, fluent.

Nevertheless, inspections might be characterized by a list of closed questions that are answered with "Yes" or "No." They are used to evaluate an institution or product according to defined requirements (e.g., premises and equipment) with the help of checklists, for instance. The inspector might look for clean benches or access restrictions in this respect.

In contrast, audits are intended to assess a process and its performance according to given standards and guidelines. They are neutral, allowing plenty of possible answers, and consist of many different observations. Audits revise theory (e.g., existing instructions) and reality according to fixed procedures with the aim to identify compliance or noncompliance with given standards, contracts, or guidelines. Accordingly, results should summarize nonconformities, which are then the subject of subsequent corrective measures. The auditor might check for documents, hierarchies, or records, for example.

Inspections and audits can be carried out, for example, with the purpose of certification (e.g., by authorities), self-control (e.g., training program), testing for the correction of deviations, and supplier or contractor qualifications. The responsible personnel should be experienced, have regulatory and technical knowledge, as well as social competences like communicative skills, chairing talents, and self-discipline, for example. Moreover, the respective employee(s) ought to support constructive cooperation, reduce potential stress, and internalize the guest appearance. It is generally possible to realize audits by internal or external personnel. The latter must be the subject of contracts and might be suitable for audits in distant countries that can be realized with the help of resident auditors.

9.1 Inspection and Audit Methodology

Inspections or audits can, for example, be realized in a system-, process-, or product-oriented way according to the following exemplified concepts:

- Trace forward (following the flow of material, starting from goods receipt to delivery)
- Product specific (starting from a completed batch documentation)
- Approval oriented (by means of approval documents aiming to compare theory and reality)
- Special reason (in case of complaints, forthcoming authority audits, corrective measures, important changes, compliance control, and much more)

However, the central subjects of inspections and audits are always related to personnel, premises, equipment, product, processes, procedures, and documentation, irrespective of which approach is employed. They should be carefully planned, and all results or findings ought to be documented. The concluding report can be used for future re-inspections, for example, with the purpose to revise the correction of former critical findings. Such activities should be timely announced, name their key points or intentions, and define their standards.

Inspectors and auditors must ensure that gained information is collected and documented. Accordingly, checklists are helpful in order not to forget essential points, to record all discussed issues, and to implement an assessment system. These lists might also provide a red line for the conduction of inspections and audits.

Commonly onsite visits are based on inspection tours and personnel interviews. People should be interviewed only during regular working time, and employees of different hierarchical level and responsibilities ought to be questioned. The interviewer should explain why this interview is done, avoid suggestive questions, be attentive, and disclose results to the interviewee. Appropriate initial question might be related to everyday work of the personnel, for example. During an inspection tour, all impressions should be communicated, documented, and uncertainties must be clarified. The tour should follow a logical and agreed upon schedule.

The audit or inspection report must be written in an understandable manner for all people involved. It should cover points such as the following:

- Date, duration, place, intention, and extent of the activities
- Summary of results including all findings, deficits, and suggested corrective measures
- Distribution list and signatures of the executing team

Findings are usually categorized as recommendations, minor, major, and critical points. Recommendations are intended to improve the quality, but they are not legally binding. Minor deviations comprise noncompliance issues that have no quality impact. Major findings do not directly affect the quality but do not comply with the regulations likewise. In contrast, critical findings might have serious effects on the product quality. Accordingly, it has to be considered that multiple findings of a lower level can be summarized to a finding of the next higher level including all consequences for the manufacturer of CBM.

9.2 Behavior during Inspections and Audits

The behavior during inspections and audits must be adapted to its importance regarding purchased materials or services and compliance with regulations, for example. Obviously, it is largely dependent on whether inspections are executed or received. Maybe the most stressful version for a manufacturer of CBM is an inspection by competent authorities that is intended to verify GMP compliance. Consequently, these events should be thoroughly prepared. This can be done via self-inspections that are carried out by employees or external auditors to show inconsistencies and for training purposes. However, official inspectors should be welcome, accompanied throughout the visit, and provided with a separate room for internal discussion and document review. Generally, only direct questions should be answered, known deviations or mistakes should not be indicated, and the quality of data or documents should not be commented upon. Furthermore, asked questions should be understood or clarified otherwise and only answered by the responsible person. No additional information or documents should be given if they were not demanded.

Nonetheless, if inspections and audits are executed, the perspective will change, but gained knowledge about the situation of being inspected should be considered. Even so, inspectors must also pay attention to some general behavioral aspects. Some of these points are exemplified and summarized as "Do's and Don'ts" and listed in Box 3.

9.3 Self-inspection

Self-inspections are mandatory according to GMP guidelines. They aim to monitor GMP compliance, identify required improvements, propose corrective measures, and control their implementation. The manufacturer has to designate responsible internal

---

Box 3 Do's and Don'ts during Execution of Audits and Inspections

Do's	Don'ts
Professional preparation and execution	Discussion among the audit team
Friendly and cooperative attitude	Excessive demands
Interest and attention for comments, answers, and descriptions	Parallel conversation, not listening, not being attentive
Objectivity and systematics	Subjectively criticizing
Incorporation of all persons concerned	Owner attitude
Appreciation of positive aspects	Work instructions for employees

or external experts who are usually found in the QC department. In case the QC department should be inspected, personnel from other departments should be employed to avoid a bias due to blind spots. Self-inspection planning is typically scheduled once a year by the QC manager, who specifies the responsible personnel and topics of inspection. The schedule should not or only minimally interfere with the activities of the inspected subjects. Common focuses of such inspections are the following:

- Personnel issues, premises, and equipment
- Production, QC, and QM system
- Implementation of former findings and corrective or improving measures
- Self-inspection and training programs

The documentation of self-inspections should consider all relevant aspects as outlined previously in this chapter.

9.4   Complaints

The manufacturer of CBM has to define respective measures if drug risks, complaints, or doubts regarding the product quality arise. These measures should be the subject of written instructions that support an appropriate and timely reaction. All complaints and other information about potential defective products must be documented in written form by a responsible, designated person. Incidents must be reported to the QP and subjected to subsequent investigations. Accordingly, retesting of affected or even further product batches might be scheduled. The handling of complaints should, for example, be focused on the following subjects:

- Clarification and description of the complaint origin
- Perpetuation of evidence and evaluation of respective documentations
- Investigation of complaint-related causality and corresponding acceptance or refusal
- Definition of CAPA procedures and complaint assessment
- Proposal of a particular reaction and external statements

All corresponding records and reports must be regularly reviewed for signs of recurring or specific problems that might necessitate product recalls. However, currently only a limited number of CBM has a market approval, and most of these products are subject of clinical trials. This implies that they are usually applied to a specific patient cohort inside the hospital that facilitates some relevant aspects like the product delivery, for example. Therefore, complaints are rare events. Nevertheless, the competent authorities have to be informed in case respective measures are envisaged by the manufacturer.

9.5   Recalls

Generally, all aspects and obligations that have been outlined for complaints are equally valid for recalls. Recalls might, for example, be required or scheduled for the following reasons:

- Deviant knowledge about certain product quality characteristics like stability data
- As a result of complaints and respective measures

- Quality defects of materials that were processed during manufacture
- Results that are obtained after release due to time requirements (e.g., sterility test)

Products that were the subject of a recall have to be labeled accordingly and must be stored separately until a further approach is decided.

9.6 Returns

Product returns might be envisaged in the case of investigational medicinal products. The GMP guidelines specify that those products that were not used during the trial have to be returned. This needs to be defined and agreed upon between the manufacturer and the sponsor. Moreover, the sponsor is responsible for the destruction of such products.

10. Conclusion

The concept of GMP is based on principles that have been developed, improved, and applied in the manufacture of pharmaceutical and health care products for decades. They intend to place the patient's safety at first rank by implementing quality standards throughout the complete life cycle of a drug. Therefore, quality and safety standards are also issued during developmental and clinical phases. In this context, the concepts of GLP and GCP shall be exemplarily named. They represent further pieces of the GxP "puzzle" and are the subject of other chapters of this book. Moreover, it is strongly recommended to consider GMP compliance during early developmental stages since delayed implementation might multiply time and cost budgets.

Nevertheless, the GMP idea was primarily focused on the pharmaceutical industry and chemical synthesizable drugs. However, the manufacture of CBM products is a raising concern. Consequently, the competent authorities are attempting to adapt the respective guidelines and regulations step-by-step. Accordingly, the manufacturer of CBM has to constantly review corresponding publications.

Nonetheless, guidelines and regulations still have to be interpreted to some extent. This is particularly valid for products that are envisaged to be applied in the sovereignty of EMA and FDA, for example. Many efforts are attempted, but comprehensive harmonization is not yet achieved.

Prospectively, it can be anticipated that regulatory requirements might become more precise, while challenges for the GMP-compliant manufacture of CBM will rise in parallel.

Acknowledgment

The author thanks Mr Daniel Kaiser and Ms Jeanine Kunz for critically reading and discussing the manuscript.

References

[1] The European Parliament and the Council of the European Union. Directive 2001/83/EC. 2007;(1394).

[2] The Innovation Office at the Paul-Ehrlich-Institut. ATMP Decision-Tree. 2011. p. 01.

[3] Maciulaitis R, D'Apote L, Buchanan A, Pioppo L, Schneider CK. Clinical development of advanced therapy medicinal products in Europe: evidence that regulators must be proactive. Mol Ther March 2012;20(3):479–82. Nature Publishing Group.

[4] European Commission. GMP guidelines Part I, Chapter 1: pharmaceutical quality system. EudraLex. 2012;4. January 2013. p. 1–8.

[5] European Commission. GMP guidelines Part I, Chapter 2: personnel. EudraLex. February 2014. p. 1–6.

[6] European Commission. GMP guidelines Part I, Chapter 3: premises and equipment. EudraLex. p. 31–4.

[7] European Commission. GMP guidelines Part I, Chapter 4: documentation. EudraLex. 2010. June 2011.

[8] European Commission. GMP guidelines Part I, Chapter 6: quality control. EudraLex. October 2014. p. 1–8.

[9] European Commission. GMP guidelines Part I, Chapter 9: self inspection. EudraLex. p. 59.

[10] European Commission. GMP guidelines Annex 16: certification by a qualified person and batch release. EudraLex. 2002;8. January 2001.

[11] International Conference on Harmonisation (ICH) of technical requirements of pharmaceuticals for human use. Guide to good manufacturing practice for medicinal products. March 2014. p. 11.

[12] U.S. Department of Health and Human Services, Food and Drug Administration, Center for Drug Evaluation and Research (CDER), Center for Biologics Evaluation and Research (CBER), Center for Veterinary Medicine (CVM) O of RA (ORA). Guidance for industry: pharmaceutical CGMP regulations. EudraLex. September 2006.

[13] European Commission. GMP guidelines Part II, Chapter 6: documentation and records. EudraLex. March 2014. p. 14–7.

[14] European Parliament. L 294/32. 2006. p. 32–50.

[15] European Parliament. Directive 2004/23/EC on setting standards of quality and safety for the donation, procurement, testing, processing, preservation, storage and distribution of human tissues and cells. 2004. p. 48–58.

[16] European Commission. GMP guidelines Annex 15: qualification and validation. EudraLex. September 2001.

[17] European Commission. GMP guidelines Part I, Chapter 5: production. EudraLex. p. 43–9.

[18] European Commission. GMP guidelines Part II, Chapter 12: validation. EudraLex. March 2014. p. 29–33.

[19] U.S. Department of Health and Human Services, Food and Drug Administration, Center for Drug Evaluation and Research (CDER), Center for Biologics Evaluation and Research (CBER) O of RA (ORA). Guidance for industry: sterile drug products current good manufacturing practice. EudraLex. September 2004.

[20] PIC/S Secretariat. Recommendations on validation master plan, installation and operational qualification, non-sterile process validation. picscheme; September 2007.

[21] PIC/S Secretariat. Recommendation of the validation of aseptic processes. picscheme; January 2011.

[22] International Conference on Harmonisation (ICH) of technical requirements of pharmaceuticals for human use. Validation of analytical procedures: text and methodology. 2005;1994. November 1996.

[23] International Conference on Harmonisation (ICH) of technical requirements of pharmaceuticals for human use. Guide to good manufacturing practice for medicinal products Part II. March 2014. p. 11.

[24] FDA. Electronic records; electronic signatures. 1997.

[25] European Commission. GMP guideline Annex 11: computerized systems. EudraLex. 2010. June 2011.

[26] European Commission. GMP guidelines Annex 1: manufacture of sterile medicinal products. EudraLex. November 2008. p. 1–16.

[27] European Commission. GMP guidelines Annex 13: investigational medicinal products. EudraLex. July 2010.

[28] European Commission. GMP guidelines Part II, Chapter 4: buildings and facilities. EudraLex. March 2014. p. 8–10.

[29] European Commission. GMP guidelines Part II, Chapter 5: process equipment. EudraLex. March 2014. p. 11–3.

[30] FDA. Subpart D: equipment and Subpart E: control of components and drug product containers and closures. Code Fed Regul 2011:157–8.

[31] FDA. Subpart C—buildings and facilities. Code Fed Regul 2011;(c):155–6.

[32] European Commission. GMP guidelines Part II, Chapter 3: personnel. EudraLex. March 2014. p. 7–8.

[33] FDA. Subpart B—organization and personnel. Code Fed Regul 2011:154–5.

[34] European Commission. GMP guidelines Part II, Chapter 8: production and in-process controls. EudraLex. March 2014. p. 20–3.

[35] European Commission. GMP guidelines Part II, Chapter 9: packaging and identification labelling of APIS and intermediates. EudraLex. March 2014. p. 23–4.

[36] European Commission. GMP guidelines Part II, Chapter 10: storage and distribution. EudraLex. March 2014. p. 25.

[37] European Commission. GMP guidelines Annex 2: manufacture of biological active substances and medicinal products for human use. EudraLex. January 2013.

[38] European Commission. GMP guidelines Annex 14: manufacture of medicinal products derived from human blood or plasma. EudraLex. 2010. November 2011. p. 1–13.

[39] U.S. Department of Health and Human Services, Food and Drug Administration, Center for Drug Evaluation and Research (CDER) C for BE and R (CBER). Guidance for industry Q7A: good manufacturing practice guidance for active pharmaceutical ingredients. August 2001.

[40] FDA. Subpart F—production and process controls. Code Fed Regul 2011:160–2.

[41] European Commission. GMP guidelines Part I, Chapter 7: outscorced activities. EudraLex. January 2013. 2012.

[42] European Commission. GMP guidelines Part I, Chapter 7: contract manufacture and analysis. EudraLex. p. 55–6.

[43] European Commission. GMP guidelines Annex 8: sampling and packaging materials. EudraLex. p. 105–6.

[44] European Commission. GMP guidelines Annex 19: reference and retention samples. EudraLex. 2006. December 2005. p. 1–5.

[45] European Commission. GMP guidelines Part II, Chapter 11: laboratory controls. EudraLex. March 2014. p. 26–9.

[46] 2.6.1. Sterility. Eur Pharmacopoeia. 2005;5.0: p. 145–9.

[47] 2.6.27. Microbiological control of cellular products. Eur Pharmacopoeia. 5.6.

[48] 2.6.7. Mycoplasmas. Eur Pharmacopoeia. 2005;5.0(1): p. 149–2.

[49] 2.6.14. Bacterial endotoxins. Eur Pharmacopoeia. 2005;5.0(01): p. 0–7.

[50] Sakudo A, Ano Y, Onodera T, Nitta K, Shintani H, Ikuta K, et al. Fundamentals of prions and their inactivation (review). Int J Mol Med [Internet] April 2011;27(4):483–9. Available from: http://www.ncbi.nlm.nih.gov/pubmed/21271212 [cited 28.01.15].

[51] Moore RA, Vorberg IPS. Species barriers in prion diseases–brief review. Arch Virol Suppl 2005;19:187–202.

[52] European Commission. GMP guidelines Part I, Chapter 8: complaints and product recalls. EudraLex. December 2005. p. 11–3.

[53] European Commission. GMP guidelines Part II, Chapter 15: complaints and recalls. EudraLex. March 2014. p. 36.

[54] European Commission. GMP guidelines Part II, Chapter 14: rejection and re-use of materials. EudraLex. March 2014. p. 34–5.

List of Abbreviations

ATMP advanced therapy medicinal products
BSC biological safety cabinets
BSE bovine spongiform encephalopathy
CAPA corrective and preventive action
CBM cell-based medicine
DQ design qualification
EDQM European Directorate for the Quality of Medicines and HealthCare
ELISpot Assay enzyme-linked immune spot assay
EMA European Medicine Agency
FACS fluorescence-activated cell sorting
FAT factory acceptance test
FRS functional requirement specification
GMP Good Manufacturing Practice
HEPA high-efficiency particulate air filter
HLA human leukocyte antigen
ICH International Conference on Harmonisation
ID identification
IPC in-process control
IQ installation qualification
LAL limulus amebocyte lysate
LPS lipopolysaccharide
OQ operational qualification
PQ performance qualification
QA quality assurance
QC quality control
QM quality management
QMH Quality Management Handbook
QP qualified person

RA risk analysis
SMF Side Master File
SOP standard operating procedure
TSE Transmissible Spongiform Encephalopathy
URS user requirement specification

Good Clinical Practice in Nonprofit Institutions

Rosario C. Mata, Ana Cardesa, Fabiola Lora
Andalusian Initiative for Advanced Therapies, Consejeria de Salud.
Junta de Andalucia, Sevilla, Spain

Chapter Outline

1. Introduction

The access of the population to a new medicinal product is usually preceded by an intense activity in basic research in animal models, in-vitro and in-vivo, and in clinical research in humans. Clinical research, as part of the biomedical research that studies prevention, diagnosis, and treatment of diseases, has as its ultimate objective to improve citizen's health, and it is also understood as being the bridge between laboratory science and clinical practice.

Clinical research is a costly and time-consuming process, often funded by a big pharmaceutical company acting as sponsor in all the different phases of a clinical research. However, in some cases, and due to some characteristics of the medicinal product, the pharmaceutical industry shows a lack of interest in their development. This is the case of the majority of cell-based medicines whose application in clinical practice can be impeded by different factors such as the difficulty of establishing a business model that could fit into the traditional model of the pharmaceutical industry. Thus, advanced therapy medicinal products (ATMPs) are frequently manufactured by small- and medium-sized biotechnological companies as well as by nonprofit public entities [1]. Therefore, it is customary that the role of sponsor usually falls either on the researchers themselves or on nonprofit institutions. This fact mainly happens in the earliest phases of the clinical research in which either academia, charity entities, or even independent researchers, all considered nonprofit institutions, lead the development of cell-based therapeutics up to early stage of clinical trials, without the resources commonly found in the big pharmaceutical or biotech industries, being, therefore, the key elements in charge of improving the health of the population.

Regardless of the means by which a new product in phase of research gets to the patient, every effort made in its development will be worthless unless the results coming from these trials are accepted by the regulatory authorities of the countries involved. Consequently, any clinical trial, no matter its phase or nature, must be done according to the specific regulatory requirements. For that reason, the codes of Good Clinical Practice (GCP) have been drawn up by most countries to ensure that the clinical trials and studies are performed in a scientific, humane, and ethical manner.

The GCP's fundamental principle is as follows: The rights, safety, and well-being of the trial subjects are the most important considerations and should prevail over interests of science and society [2].

With this aim in mind, a balance should be found to ensure that an overprotective approach does not hinder the access to some particular cell therapy products. It is essential to take into account the risk–benefit balance when setting out quality requirements in cell products. In those in which no substantial manipulation is carried out, and, therefore, being borderline products between medicinal product and cell transplant, to demand those equal requirements claimed for traditional medicines may represent an enormous obstacle to overcome for their application, delaying or impeding the access to possible safe and effective therapies to patients with diseases lacking any other therapeutic alternative [3].

Despite making reference to "Good Clinical Practice in nonprofit institutions," the title of this chapter, it should be highlighted that there is not any difference between a for-profit and a nonprofit organization when it comes to considering how the codes of GCP must be applied during the development of a clinical research. However, what seems clear is that the compliance with these codes may be more or less difficult depending on the resources you may have available to carry out the research.

When we talk about nonprofit institutions sponsoring a clinical research, we should distinguish different scenarios.

In some cases, an independent researcher, just on his/her own, decides to start up a clinical research in its early stage of development and, consequently, acts as the sponsor as well as the principal investigator along with the support of the health institution where the clinical research is being taken place as it usually happens in some European countries. This fact does not exempt him from all the responsibilities arising from each of the roles he/she is assuming, that is, from the search for funding to guarantee the supply of medicine, the design and preparation of all essential documents related to the clinical trial, their processing with competent authorities, etc. However, this scenario is unusual when it comes to developing an ATMP, due mainly to the difficulty added by the specific existing regulation for this kind of product and, specially, by the limited access to laboratories that manufacture medicinal products under Good Manufacturing Practice (GMP) conditions.

A second more advantageous scenario is that in which, within the public health system, some organizations are established to sponsor the development of new therapies to improve the population's health and to incorporate innovative advanced therapies in the health care and progress of the region.

Examples of this alternatives are the UK Cell Therapy Catapult (https://ct.catapult. org.uk/) established in 2012 or the strategy founded by the Andalusian government (a region in the south of Spain), known as The Andalusian Initiative for Advanced Therapies (AIAT) [4,5], which was established to sponsor the development of new therapies to improve the population's health and to incorporate innovative advanced therapies in the health care and for the progress of the region. To do this, alliances with the academic world, research institutions, health centers, patients' associations, small and medium enterprises, and the pharmaceutical industry must be sought [6]. AIAT has created a platform for technology maturation for investigational advanced therapy products, supporting translational research from the preclinical development stages with a special focus on clinical research. This type of organization, frequently public foundations, provides not only logistical, methodological, and regulatory support to researchers but also the facilities where clinical trials can be conducted, all along their different stages, guaranteeing their development under GCP. In addition, they also support the researchers in their search for competitive funding sources.

Does this difference in resources affect the GCP and the final results of a clinical research?

It would be reasonable to believe that the development of a medicinal product when promoted, during its clinical research stages, by the pharmaceutical industry whose resources, not only financial but also human, are more abundant would guarantee a higher quality in results. However, empirical evidence shows that conclusions in randomized trials are significantly more positive toward the experimental interventions in trials funded by for-profit organizations than those trials lacking competing interest, due to biased interpretation of trials results [7]. Recent controversies over the protection of human subjects, payment of physicians for recruiting patients to clinical trial, Food and Drug Administration (FDA) removal of approved drugs from the market, and reporting of results of clinical trials have highlighted important facets of clinical research.

1.1 History of GCP Legislation

GCP is based on the ethical principles and subjects' rights in research, which are protected according to several publicized guidelines, such as the Nuremberg Code [8], the Belmont Report [9], and the Declaration of Helsinki [10].

Up to the beginning of the twentieth century, the medicine was lacking well-established therapeutic methods and the advance in the knowledge was essentially empirical. At the outset of the twentieth century, the "scientific medicine" was developed, as well as, later on, in its second half, what was named the "therapeutic explosion" with the appearance of a multitude of numerous medicines, antibiotics being the first.

However, the drafting of basic criteria regulating the ethical aspects of the clinical research was only completed in the last decade and, in many cases, they were reached as a result of the response to dramatic situations coming from experiments that caused unnecessary pain and distress to many human beings.

In the last decades of the twentieth century, diverse procedures or obligations were published by different authorities, as the American one, the first European Directives, and the GCP Recommendations of the World Health Organization (WHO).

Thus, in 1995, during the International Conference of Harmonization, also known as the International Conference on Harmonisation or ICH [2], the European Union (EU), Japan, and the United States agreed to develop a common guide with regard to GCP. Clinical trials had to fulfill it when submitted to get the authorization of medicinal products in the mentioned territories. The document was approved in 1996 by the Committee of Medicines of Human Use (CHMP) of the European Agency of Medicines (European Medicines Agency, EMA), came into force in 1997, and provided the EU, the United States, and Japan with a unified frame.

In that document, entitled Note for Guidance on Good Clinical Practice, CPMP/ ICH/135/95 [11], GCP is defined as an international standard with scientific and ethical quality directed toward the design, registry, and writing of reports on clinical trials involving humans. The fulfillment of these standards policies guarantees the protection of the rights, safety, and well-being of the subjects participating in the trial and ensures the credibility of the data collected in a clinical trial.

Subsequently, the European Directive 2001/20/EC [12], the norm nowadays repealed by the Regulation EU No 536/2014 of the European Parliament and of the Council of April 16, 2014 [13], established the legal and administrative basis for the application of the GCP standards in clinical trials with medicinal products conducted in Europe. This Directive was transposed in the different countries of the EU.

Years later, Directive 2005/28/EC [14] implemented the principles and detailed guidelines of GCP in relation to investigational medicinal products (IMPs) for human use, as well as the requirements to authorize the manufacturing or importation of these medicines.

For ATMPs, the guidelines "Detailed guidelines on good clinical practice specific to advanced therapy medicinal products" [15], published for the European Commission in 2009, supplement the principles and detailed guidelines set out in the Commission Directive 2005/28/EC of April 8, 2005, laying down principles and detailed guidelines for GCP as regards IMPs for human use, as well as the requirements for authorization of the manufacturing or importation of such products. They should be read in conjunction with the detailed guidelines set out in Volume 10 of the Rules Governing Medicinal Products in the European Union, including in particular the Note for Guidance on Good Clinical Practice, as well as other guidelines specific to advanced therapies.

The clinical trials, in which the marketing authorization of medicinal products in Europe is based, must follow GCP in accordance with Directive 2004/27/EC [16].

The Regulation (EU) No 536/2014 [13] of the European Parliament and of the Council, of April 16, 2014, on clinical trials on medicinal products for human use, establish, in its article 2, paragraph 30, the definition of the term clinical practice, meaning a set of detailed ethical and scientific quality requirements for designing, conducting, performing, monitoring, auditing, recording, analyzing, and reporting clinical trials ensuring that the rights, safety, and the well-being of the subjects are protected, and that the data generated in the clinical trial are reliable and robust.

The people involved in a clinical trial must not only consider GCP standards, but also scientific guidelines on quality, safety, and efficacy of medicinal products for human use adopted by the CHMP [17] and published by the EMA, as well as the other communitarian pharmaceutical guidelines published by the European Commission in the different volumes of the Standards on Medicinal Products in the European Union.

The ICH guidelines documents, as with regional agency guidance documents (e.g., FDA), are considered nonbinding recommendations and as such do not represent legally enforceable responsibilities. However, these guidance documents describe any agency's current thinking and are viewed as recommendations unless they are transposed to the national laws of different countries.

Probably the first warning about the aspects that were necessary to bear in mind at the moment of performing a clinical trial, with human beings, were the judgments of Nuremberg where there were judged the crimes perpetrated by the Nazi's doctors in concentration camps during the Second World War.

1.1.1 Nuremberg Code

The Nuremberg Code on medical ethics sets a series of principles for human experimentation as a result of the Nuremberg Trials at the end of the Second World War [8]. The Code responds to the deliberations and arguments by which the Nazi hierarchy and some physicians were tried due to the cruel treatment given to the prisoners of concentration camps.

The Nuremberg Code was published on August 20, 1947, after the Nuremberg Trials, held between August 1945 and October 1946. The Code establishes the principles for human experimentation, considering that several of the accused argued during the trials that their experiments differed little from those performed before the war due to a lack of legislation on human experimentation.

The Ten Points of the Nuremberg Code	
1	Consent must be voluntary
2	The results cannot be procured by other methods
3	Research must be based on animal experimentation
4	Unnecessary suffering and injury should be avoided
5	Research causing the death or injury of a subject is illegitimate
6	Benefits should exceed risks
7	Measures to protect the subjects should be taken
8	The experiment should be conducted by qualified personnel
9	If desired, the subject can bring the study to an end
10	The scientist in charge must be prepared to terminate the study at any stage

However, these 10 points were never granted legal recognition, nor did they have an impact on medical publications. It was not until 1964 that they were taken into consideration in the Declaration of Helsinki, approved by the World Medical Association (WMA) in 1964 [10].

1.1.2 Kefauver Harris Amendment

Thalidomide is a drug that was commercialized in Europe between 1958 and 1963 as a sedative and as a tranquilizer used to reduce nausea during the first three months of pregnancy. Although it was not authorized in the United States, it was still used and patients were not informed that it was not authorized.

The drug caused the so-called "thalidomide disaster," in which thousands of babies worldwide were born with severe irreversible malformations. In fact, its teratogenicity would never have been known if the malformation that it caused had been more common.

The Kefauver Harris Amendment [18] was a reply to the thalidomide tragedy. In the United States, Doctor Frances Oldham Kelsey, the supervisor of the FDA (the North American regulatory agency equivalent to the EMA in Europe) refused to authorize the drug for the market because she had concerns about its safety. The US citizens became outraged, and in response, Senator Estes Kefauver from Tennessee and Representative Oren Harris from Arkansas presented a draft requiring drug manufacturers to prove the efficacy and safety of their products prior to their approval.

An "effectiveness test" was introduced, a requirement that was not present previously. In addition, the amendment required drug advertisements to compulsorily include accurate information about any possible side effects and the effectiveness of the treatments. Likewise, it was required that cheap generic drugs no longer be marketed as high-cost drugs with new commercial names and introduced as a new "breakthrough" in pharmacology, as was done prior to the amendment. The law was signed by President John F. Kennedy on October 10, 1962.

1.1.3   The Tuskegee Syphilis Experiment

The Tuskegee Syphilis Experiment [19] was a clinical study conducted between 1932 and 1972 in Tuskegee, Alabama, by the United States Public Health Service. Four hundred Afro-American sharecroppers, most of them illiterate, were studied to observe the natural progression of untreated syphilis up to their eventual death by the disease.

This experiment aroused controversy and led to changes in the legal protection of the patients involved in clinical studies. Subjects involved in this experiment did not give their informed consent; they were not informed of their diagnosis and were told that they were being treated for "bad blood." They were also told that if they participated in the study, they would be given free medical care, free transportation to the clinic, free meals, and free burial insurance in the case of death. Subjects were also warned to avoid penicillin treatment, which was already in use with other patients nearby.

In 1932, when the study began, treatments for syphilis were very toxic, dangerous, and had questionable effectiveness. Part of the purpose of the study was to determine if the benefits of the treatment compensated its toxicity and to recognize the different stages of the disease to develop treatments adapted to each of those stages. Doctors recruited 399 black men, supposedly infected with syphilis, to study the progress of the disease for the 40 following years. A control group of 201 healthy men was also studied to establish comparisons.

In 1947, penicillin had become the treatment of choice for syphilis. Before this finding, syphilis frequently led to a chronic, painful disease, and it eventually caused multiple organ failure. Instead of treating the subjects of the study with penicillin and concluding it or establishing a control group to study the drug, the scientists in charge of the Tuskegee experiment hid the information on penicillin from the subjects in order to continue studying how the disease spread and eventually led to death.

The study continued until 1972 when it was leaked to the press, thus bringing it to an end. By then, 28 of the 399 patients had died from syphilis and another 100 from related medical complications. In addition, 40 patients' wives were infected and 19 children contracted the disease when being born.

The Tuskegee experiment led to the Belmont Report [9] of 1979 and the creation of the National Human Investigation Board, as well as the request for the creation of institutional review boards (IRBs).

1.1.4   Declaration of Helsinki

The Declaration of Helsinki [10] was published by the WMA as a set of ethical principles that were established to guide the medical community and any other people working in human subject research. Despite the fact that it is not a legally binding instrument under the international law, it is widely considered as the most important document on human subject research ethics. It draws its authority from the degree to which it has been codified internally and from the influence it has gained nationally and internationally.

The Declaration was originally adopted in June 1964 in Helsinki, Finland, and has since undergone seven revisions and two clarifications, thus leading to its extension from 11 to 32 paragraphs (the last published update is from October 2013). The Declaration is an important document in the history of research ethics as a result of the significant efforts by the medical community to develop regulations. It is also the basis for many subsequent documents.

The previous Nuremberg Code did not gain general acceptance on the ethical aspects of human research, although it inspired national policies on human research in countries like Germany and Russia. The Declaration of Helsinki broadens the 10 points of the Nuremberg Code, which is added to the Declaration of Geneva (1948), a declaration on the ethical duties of physicians. The Declaration of Helsinki addresses clinical research in detail and reflects changes in medical practice even from the term "human experimentation" used in the Nuremberg Code.

The basic principle of the Declaration of Helsinki is respect for the individual, their right to self-determination, and the right to make informed decisions (informed consent) regarding participation in the investigation, both initially and during the course of the investigation. The investigators' duty is solely toward the patient or volunteer, and while there is always a need for research, the subject's well-being must always prevail over the interests of science or society. Likewise, ethical considerations must always come from a previous analysis of laws and regulations.

The recognition of the increasing vulnerability of individuals and groups requires special vigilance. It is acknowledged that when the participant is incompetent, physically or mentally incapable of giving consent, or is a minor, allowance should be considered for surrogate consent by an individual acting in the subject's best interest.

1.2   The Principles of ICH GCP

Clinical trials should be conducted in accordance with the ethical principles that have their origin in the Declaration of Helsinki, and that are consistent with GCP and the applicable regulatory requirement(s).

Before a trial is initiated, foreseeable risks and inconveniences should be weighed against the anticipated benefit for the individual trial subject and society. A trial should be initiated and continued only if the anticipated benefits justify the risks.

The rights, safety, and well-being of the subjects participating in the trial are the most important considerations and should prevail over interests of science and society.

The available nonclinical and clinical information on an investigational product should be adequate to support the proposed clinical trial.

Clinical trials should be scientifically sound, and described in a clear, detailed protocol.

A trial should be conducted in compliance with the protocol that has received prior IRB/independent ethics committee (IEC) approval/favorable opinion.

The medical care given to, and medical decisions made on behalf of, subjects should always be the responsibility of a qualified physician or, when appropriate, of a qualified dentist.

Each individual involved in conducting a trial should be qualified by education, training, and experience to perform his or her respective task(s).

Freely given informed consent should be obtained from every subject prior to clinical trial participation.

All clinical trial information should be recorded, handled, and stored in a way that allows its accurate reporting, interpretation, and verification.

The confidentiality of records that could identify subjects should be protected, respecting the privacy and confidentiality rules in accordance with the applicable regulatory requirement(s).

Investigational products should be manufactured, handled, and stored in accordance with applicable Good Manufacturing Practice (GMP). They should be used in accordance with the approved protocol.

Systems with procedures that assure the quality of every aspect of the trial should be implemented.

2. The Elements of GCP Compliance

To ensure a proper compliance with GCP, the involvement of three elements, as seen in Figure 1, is entirely necessary: Sponsor, Investigator, and the IRB. The supervision of the clinical trial is a responsibility shared by these three elements.

The investigator and his/her team should know that the sponsor has overall responsibility for the proper conduct of the trial. However, some investigators and their

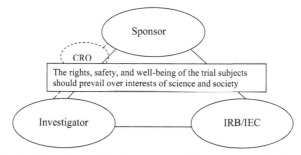

Figure 1 The clinical trial triangle. The three main entities involve in the clinical trials.

nonprofit institutions, developing clinical trials, may also act as sponsor. Therefore, it is essential for those investigators and their institutions to know and understand all the implied responsibilities of the role of sponsor.

The sponsor shall delegate any or all trial-related functions to a clinical research associate (CRA) or a contract research organization (CRO). However, the sponsor shall remain responsible for ensuring that the conduct of the trials and the final data generated in these trials conform to the guidance GCP. The investigator and the sponsor may be the same person.

2.1 The Sponsor: Duties and Responsibilities

The ICH Guideline [11] defines the Sponsor as an individual, company, institution, or organization that takes responsibility for the initiation, management, and/ or financing of a clinical trial. The duties and responsibilities can be summarized as follows.

2.1.1 Quality Assurance and Quality Control

The sponsor is responsible for implementing and maintaining quality assurance and quality control systems with written standard operating procedures (SOPs) to ensure that trials are conducted and data are generated, documented (recorded), and reported in compliance with the protocol, GCP, and the applicable regulatory requirement(s).

2.1.2 Contract Research Organization

The sponsor may delegate any or all of their responsibilities to a CRO, but the ultimate responsibility for the quality and integrity of the trial data always resides with the sponsor. In these cases, the specific responsibilities, duties, and functions of the CRO should be written in a contract.

2.1.3 Medical Expertise

The sponsor should designate appropriately qualified medical staff that will be capable to advise on trial-related medical questions, review documents, or resolve problems. This task can be transferred to the chief investigator, or it can be contracted if the sponsor does not have suitable medical staff in-house.

2.1.4 Trial Design, Management, Data Handling, and Record Keeping

The sponsor should use qualified people (e.g., biostatisticians, clinical pharmacologists, and physicians) as appropriate, throughout all stages of the trial process.

The management of the clinical trial by the sponsor will involve the use of qualified people to oversee the development, management, and verification of the data, to perform statistical analyses, and elaborate on the final clinical trial reports. There are ICH

Guides for the Protocol design and other ICH guides on the Structure and Content of Clinical Trial reports [20].

When using electronic trial data handling and/or remote electronic trial data systems, the sponsor should ensure and document that the electronic data processing system(s) conforms to the sponsor's established requirements for completeness, accuracy, reliability, and consistent intended performance (i.e., validation).

2.1.5   Investigator Selection

The sponsor is responsible for selecting the investigator(s) and their institution(s). Each investigator should be qualified by training and experience and should have adequate resources to properly conduct the trial for which the investigator is selected.

Likewise, the sponsor is responsible for verifying that the sites/facilities where the trial will be conducted are appropriate. Thus, the sponsor can make a "feasibility visit" to select the center (it is not mandatory, but is recommended) in which a professional designated by the sponsor, generally a monitor, carries out the following activities: interviews with the principal investigator (PI), visually checks the site to see if it has the resources needed, and visits the site's pharmacy or the place where the medicinal products will be received, stored, and protected.

2.1.6   Compensation to Subjects and Investigators

The sponsor will provide financial compensation to subjects in case of trial-related injury or death. The sponsor will provide the investigator with the legal and financial coverage in these cases except when the injury is the result of the investigator's negligence or malpractice.

2.1.7   Notification/Submission to Regulatory Authority(s)

Before initiating the clinical trial(s), the sponsor should submit any required application(s) to the appropriate authority(s) for review, acceptance, and/or permission (as required by the applicable regulatory requirement(s)) to begin the trial(s). Any notification/submission should be dated and contain sufficient information to identify the protocol.

2.1.8   Confirmation of Review by IRB/IEC

The sponsor should obtain the IRB/IEC approval/favorable opinion.

If the IRB/IEC conditions its approval/favorable opinion upon change(s) in any aspect of the trial, such as modification(s) of the protocol, written informed consent form, and any other written information to be provided to subjects, and/or other procedures, the sponsor should obtain from the investigator/institution a copy of the modification(s) made and the date approval/favorable opinion was given by the IRB/IEC. The sponsor should obtain from the investigator/institution documentation and dates of any IRB/IEC reapprovals/re-evaluations with favorable opinion, and of any withdrawals or suspensions of approval/favorable opinion.

2.1.9 Investigational Medicinal Product

One of the major responsibilities of the sponsor is related to the investigational medication, such as the manufacturer, labeling, storage, handling, and distribution. Manufacturer must be according to the requirements of GMP.

Therefore, the sponsor is responsible for supplying the investigator(s)/institution(s) with the investigational product(s), always after obtaining all required documentation (e.g., approval/favorable opinion from IRB/IEC and regulatory authority(s)).

In addition, the sponsor should ensure that written procedures include instructions that the investigator/institution should follow for the handling and storage of investigational product(s) for the trial and documentation thereof. The procedures should address adequate and safe receipt, handling, storage, dispensing, retrieval of unused product from subjects, and return of unused investigational product(s) to the sponsor (or alternative disposition if authorized by the sponsor and in compliance with the applicable regulatory requirement(s)).

In blinded trials, the coding system for the investigational product(s) should include a mechanism that permits rapid identification of the product(s) in case of a medical emergency, but does not permit undetectable breaks of the blinding.

The sponsor should update the Investigator's Brochure (IB) when significant new information becomes available.

2.1.10 Adverse Drug Reaction Reporting

The sponsor is responsible for communicating any suspicions of serious or unexpected adverse reactions to the regulatory authorities, the investigators, and the IRB/IEC involved in the study. In the same way, the investigational product must undergo a continuous evaluation in terms of safety.

2.1.11 Monitoring, Audits, and Inspections

All the information related to the clinical trial must be recorded, handled, and filed so that accurate reporting, interpretation, and verification can be carried out. For that purpose, GCP establishes three methods for control and quality assurance: clinical trial monitoring, audits, and inspections.

2.1.12 Premature Termination or Suspension of a Trial

If there is a premature conclusion/suspension of a trial, the sponsor must inform the investigators/institutions and the regulatory authorities in a timely manner of the reasons that caused it.

2.2 The Clinical Investigator: Duties and Responsibilities

According to the GCP [11], the clinical investigator is a person responsible for the conduct of the clinical trial at a trial site. If a trial is conducted by a team of individuals at a trial site, the investigator is the leader responsible for the team and may be called the PI. The main responsibilities of the clinical investigator are, among others, the following.

2.2.1 Investigator's Qualifications and Agreements

The investigator should be qualified by education, training, and experience to assume responsibility for the proper conduct of the trial, should meet all the qualifications specified by the applicable regulatory requirement(s), and should provide evidence of such qualifications through up-to-date curriculum vitae (CV) and/or other relevant documentation requested by the sponsor, the IRB/IEC, and/or the regulatory authority(s).

2.2.2 Communication with IRB/IEC

Before starting a trial, the investigator/institution should have written and dated approval/favorable opinion from the IRB/IEC for the trial protocol, a written informed consent form, consent form updates, subject recruitment procedures (e.g., advertisements), and any other written information to be provided to subjects.

2.2.3 Compliance with Protocol

The investigator must strictly follow the protocol while conducting the clinical trial. To implement it, the investigator and the sponsor must agree and sign the trial protocol.

The investigator shall not implement any deviations to the protocol without the previous approval of the sponsor. However, there is an exception: if a patient's life is in danger, the investigator will firstly guarantee the well-being of the patient.

Sometimes, it is necessary to modify or amend a clinical trial protocol. If this happens, the investigator must follow the amendment of the protocol strictly once it has been approved when performing the rest of the clinical trial. To implement, it the investigator and the sponsor must agree and sign the amendment.

2.2.4 Investigational Product(s)

Investigators should be thoroughly familiar with the properties and use of the IMP, as described in the IB. The investigator must document the reading of the IB and the knowledge of risks and potential adverse reactions related to the IMP.

In addition, the sponsor must inform the investigator about all relevant new data about the product appearing during the course of the trial.

The investigator's responsibilities are the receipt, storage, dispensing, use, and disposal; however, these accountabilities are assigned to the pharmacist of the institution.

2.2.5 Informed Consent of Trial Subjects

The investigator is responsible for providing the subjects with accurate information about the clinical trial, what it consists of, and its risks and benefits. The investigator should adhere to the current version of the Declaration of Helsinki. The subjects must be informed verbally and in writing (they will be given a copy of the patient information and the informed consent form), and the language used must be understandable to the subject. Neither the investigator nor the trial staff should coerce or unduly influence a subject to participate or to continue to participate in a trial. If a subject is unable to read or if a legally acceptable representative is unable to read, an impartial witness should be present during the entire informed consent discussion.

It is important to highlight that a subject can withdraw from the trial at any time, without having to explain the reasons for doing so and without penalty or loss of benefits concerning the current or future medical care they are entitled to.

The investigator will also provide subjects with information about the person(s) to be contacted in case of emergency and urge subjects to carry this information with them.

In emergency situations, when prior consent of the subject is not possible, the consent of the subject's legally acceptable representative, if present, should be requested. When prior consent of the subject is not possible, and the subject's legally acceptable representative is not available, enrollment of the subject should require measures described in the protocol and/or elsewhere, with documented approval/favorable opinion by the IRB/IEC, to protect the rights, safety, and well-being of the subject and to ensure compliance with applicable regulatory requirements. The subject or the subject's legally acceptable representative should be informed about the trial as soon as possible and consent to continue, and other consent as appropriate should be requested.

2.2.6   Records and Reports

The investigator should ensure the accuracy, completeness, legibility, and timeliness of the data reported to the sponsor in the case report form (CRF) and in all required reports. The investigator will complete the CRF with the recorded data in the patient's medical record, and not vice versa. The CRF must be completed, but it will never be a source document and will never replace the medical records.

2.2.7   Safety Reporting

The investigator must immediately notify the sponsor of any serious or unexpected adverse events.

2.3   The Monitor: Duties and Responsibilities

The objective of the monitoring processes is to protect the rights, safety, and welfare of subjects participating in clinical trials.

As we mentioned previously, a sponsor can transfer some or all trial-related duties or functions to a monitor or a CRO, but the sponsor is ultimately responsible for the quality and integrity of the trial data. The monitor can be an employee of the Sponsor or CRO.

The monitor may be physicians, veterinarians, nurses, pharmacist, etc. The monitor's qualifications shall be documented in their updated CV, copies of their academic qualifications, training certificates, and other training-related records.

Monitors must be thorough with the IMP, the protocol, the informed consent form, and any other written information provided to the subjects. They should also be familiar with the SOPs, GCPs, and the legislation.

The monitor's responsibilities are described next.

2.3.1 Selection Visits

This visit, which is not mandatory but is recommended, usually takes place in a very preliminary phase of the clinical trial. It is focused on learning about the site's recruitment capabilities and performing the specific tests of the clinical trial protocol.

2.3.2 Initial Visits

The purpose of this visit is to document that all clinical trial procedures have been checked with the PI and the collaborative staff to ensure their correct implementation and to ensure that the site has all the necessary resources and that is ready to begin with the recruitment of patients of the clinical trial.

The following are the points to be considered in an initial visit:

Understands the nature of the protocol
The investigator team is adequately experienced and qualified to perform the study
Understands the procedure to obtain informed consent in accordance with GCP
The investigator team has the access to a number of patients for inclusion in the clinical trial and has sufficient time to carry out the responsibilities with the trail.
Recording and notification of adverse events

2.3.3 Visits During the Trial

There is a document called the monitoring plan, which describes the type of monitoring to be carried out, its planning, minimum content, and reporting mechanisms. The frequency of the monitoring visits is described in this document.

Objectives:

- To verify compliance with the protocol and GCP standards by proposing corrective action to deviations (if applicable)
- To check the integrity, veracity, accuracy, and traceability of the data that are recorded in the CRF against the source documents
- To solve the research team's doubts and problems throughout the trial
- To check the number of evaluated patients including those who finished the trial and those who withdrew from it
- To trace the investigational product from its reception to its dispensing and destruction
- To verify the obtaining of the informed consent documents
- To review the records and reports of adverse events
- To verify that the CRFs are completed and corrected properly
- To review and update the investigator's file
- To check that the source documents register the progress of the clinical trial correctly

2.3.4 Closing Visits

The close-out visit is performed to bring the trial to a proper close at the trial site. The aim is to clarify any open questions about the data collected, to ensure that any remaining IMPs, medical devices, and trial materials remaining are dealt with correctly, and to discuss responsibilities after the end of the trial.

Objectives:

- To ensure that all the pages of the CRFs have been collected and that all inconsistencies and queries have been solved
- To ensure that the monitoring documents of all adverse events have been collected
- To ensure that there are no corrective actions pending
- To ensure that the investigator's file contains all the essential documents
- To ensure that all the product and the surplus materials have been destroyed/returned
- To ensure that the PI is informed of their responsibilities once the trial closes

2.3.5 Monitoring Report

The monitor should submit a written report to the sponsor after each trial site visit or trial-related communication. The objective is to report about the visit and give specific details about the information reviewed, the findings, deviations, and relevant deficiencies, as well as the conclusions, actions taken or planned, and/or recommended actions with the aim of guaranteeing compliance.

2.4 The Institutional Review Board: Independent Ethics Committee

The IRB is an independent body constituted by medical, scientific, and nonscientific members, whose responsibility is to ensure the protection of the rights, safety, and well-being of human subjects involved in a trial by, among other things, reviewing, approving, and providing continuing review of trial protocol and amendments and of the methods and material to be used in obtaining and documenting informed consent of the trial subjects.

One of the principles of ICH GCP is to obtain the approval/favorable opinion by the IRB/IEC. The application of the ethical review should be submitted by the sponsor/investigator. Thus, the IRB/IEC should obtain and review the following documents:

Trial protocol(s)/amendment(s)
Written informed consent form(s) and consent form
Updates that the investigator proposes for use in the trial
Subject recruitment procedures (e.g., advertisements), written information to be provided to subjects
Investigator's Brochure
Information about payments and compensation available to subjects
The investigator's current CV and/or other documentation evidencing qualifications

2.4.1 Composition

The IRB should consist at least of five members; all of them should be professionally qualified to review the study documents. At least one member shall be one whose primary area of interest is in a nonscientific area, and another should be independent of the institution/trial site.

The IRB/IEC must safeguard the rights, safety, and well-being of all trial subjects. Primarily, they shall pay special attention to studies involving subjects who may be more vulnerable, such as children, pregnant women, or mentally disabled persons.

3. The Clinical Trial Protocol

A clinical trial protocol is a document that describes the objective(s), design, methodology, statistical considerations, the ethical and legal issues, and the practical organization of a clinical trial. The protocol should normally also provide the background and reasons why the trial is being conducted.

The protocol contains the study plan on which the clinical trial is based. The plan is designed to safeguard the health of the participants as well as answer specific research questions.

The format and content of clinical trial protocols sponsored by nonprofit institutions, in the same way that those promoted for pharmaceutical, biotechnology, or medical device companies, has been standardized and should follow GCP guidance issued by the International Conference on Harmonization of Technical Requirements for Registration of Pharmaceuticals for Human Use [11].

In the EU, several Directives have been published to define guidelines and practices for the design and carrying out of drug clinical trials.

The existence of a clinical trial protocol allows researchers at multiple locations (in case of a multicenter trial) to perform the study in exactly the same way.

The contents of a trial protocol should generally include the following topics.

3.1   General Information

Protocol title, protocol identifying number, and date
Name and address of the sponsor and monitor (if other than the sponsor)
Name and title of the person(s) authorized to sign the protocol
Name, title, address, and telephone number of the sponsor's medical expert for the trial
Name and title of the investigator(s) responsible for the conduct of the study, with address and telephone number of each trial site
Name, title, address, and telephone number of any other qualified physician responsible for trial-related medical decisions, if not the investigator
Names and addresses of all clinical laboratories and other medical and/or technical departments or institutions involved in the study

3.2   Background Information

Name and description of the investigational product(s)
A summary of findings from nonclinical studies that potentially have clinical significance and from clinical trials that are relevant to the trial
Summary of the known and potential risks and benefits, if any, to human subjects

Description of and justification for the route of administration, dosage, dosage regimen, and treatment period(s)

A statement that the trial will be conducted in compliance with the protocol, GCP, and the applicable regulatory requirement(s)

Description of the population to be studied

References to literature and data that are relevant to the trial, and that provide background for the trial

3.3 Trial Objectives and Purpose

• A detailed description of the objectives and the purpose of the trial

3.4 Trial Design

The scientific integrity of the trial and the credibility of the data from the trial depend substantially on the trial design. A description of the trial design should include the following:

A specific statement of the primary endpoints and the secondary endpoints, if any, to be measured during the trial

A description of the type/design of trial to be conducted (e.g., double-blind, placebo-controlled, parallel design) and a schematic diagram of trial design, procedures, and stages

A description of the measures taken to minimize/avoid bias, including:

Randomization
Blinding

A description of the trial treatment(s) and the dosage and dosage regimen of the investigational product(s). Also include a description of the dosage form, packaging, and labeling of the investigational product(s).

The expected duration of subject participation, and a description of the sequence and duration of all trial periods, including follow-up, if any.

A description of the "stopping rules" or "discontinuation criteria" for individual subjects, parts of trial, and entire trial.

Accountability procedures for the investigational product(s), including the placebo(s), and comparator(s), if any.

Maintenance of trial treatment randomization codes and procedures for breaking codes.

The identification of any data to be recorded directly on the CRFs (i.e., no prior written or electronic record of data), and to be considered to be source data.

3.5 Selection and Withdrawal of Subjects

Subject inclusion criteria
Subject exclusion criteria
Subject withdrawal criteria (i.e., terminating investigational product treatment/trial treatment) and procedures specifying the following:
a. When and how to withdraw subjects from the trial/investigational product treatment
b. The type and timing of the data to be collected for withdrawn subjects

c. Whether and how subjects are to be replaced

d. The follow-up for subjects withdrawn from investigational product treatment/trial treatment

3.6   Treatment of Subjects

The treatment(s) to be administered, including the name(s) of all the product(s), the dose(s), the dosing schedule(s), the route/mode(s) of administration, and the treatment period(s), including the follow-up period(s) for subjects for each investigational product treatment/trial treatment group/arm of the trial

Medication(s)/treatment(s) permitted (including rescue medication) and not permitted before and/or during the trial.

Procedures for monitoring subject compliance

3.7   Assessment of Efficacy

Specification of the efficacy parameters

Methods and timing for assessing, recording, and analyzing of efficacy parameters

3.8   Assessment of Safety

Specification of safety parameters

The methods and timing for assessing, recording, and analyzing safety parameters

Procedures for eliciting reports of and for recording and reporting adverse event and intercurrent illnesses

The type and duration of the follow-up of subjects after adverse events

3.9   Statistics

A description of the statistical methods to be employed, including timing of any planned interim analysis

The number of subjects planned to be enrolled. In multicenter trials, the numbers of enrolled subjects projected for each trial site should be specified. The reason for choice of sample size, including reflections on (or calculations of) the power of the trial and clinical justification

The level of significance to be used

Criteria for the termination of the trial

Procedure for accounting for missing, unused, and spurious data

Procedures for reporting any deviation(s) from the original statistical plan (any deviation(s) from the original statistical plan should be described and justified in protocol and/or in the final report, as appropriate)

The selection of subjects to be included in the analyses (e.g., all randomized subjects, all dosed subjects, all eligible subjects, evaluable subjects)

3.10   Direct Access to Source Data/Documents

The sponsor should ensure that it is specified in the protocol or other written agreement that the investigator(s)/institution(s) will permit trial-related monitoring, audits, IRB/IEC review, and regulatory inspection(s), providing direct access to source data/documents.

3.11   Quality Control and Quality Assurance

3.12   Ethics

Description of ethical considerations related to the trial

3.13   Data Handling and Record Keeping

3.14   Financing and Insurance

Financing and insurance if not addressed in a separate agreement

3.15   Publication Policy

Publication policy, if not addressed in a separate agreement

3.16   Reports

Since the protocol and the clinical trial/study report are closely related, further relevant information can be found in the ICH Guideline for Structure and Content of Clinical Study Reports [20].

4.   The Investigator's Brochure

The IB is a compilation of the clinical and nonclinical data on the investigational product(s) that are relevant to the study of the product(s) in human subjects [8]. Its purpose is to provide the investigators and others involved in the trial with the information to facilitate their understanding of the rationale for, and their compliance with, many key features of the protocol, such as the dose, dose frequency/interval, methods of administration, and safety monitoring procedures. The IB also provides insight to support the clinical management of the study subjects during the course of the clinical trial. The information should be presented in a concise, simple, objective, balanced, and nonpromotional form that enables a clinician, or potential investigator, to understand it and make his/her own unbiased risk–benefit assessment of the appropriateness of the proposed trial. For this reason, a medically qualified person should generally participate in the editing of an IB, but the contents of the IB should be approved by the disciplines that generated the described data.

It is expected that the type and extent of information available will vary with the stage of development of the investigational product. If the investigational product is marketed and its pharmacology is widely understood by medical practitioners, an extensive IB may not be necessary. Where permitted by regulatory authorities, a basic product information brochure, package leaflet, or labeling may be an appropriate alternative, provided that it includes current, comprehensive, and detailed information on all aspects of the investigational product that might be of importance to the investigator. If a marketed product is being studied for a new use (i.e., a new indication), an IB specific to that new use should be prepared. The IB should be reviewed at least annually and revised as necessary in compliance with a sponsor's written procedures. More frequent revision may be appropriate depending on the stage of development and the generation of relevant new information. However, in accordance with GCP, relevant new information may be so important that it should be communicated to the investigators, and possibly to the IRBs/IECs and/or regulatory authorities before it is included in a revised IB.

Generally, the sponsor is responsible for ensuring that an up-to-date IB is made available to the investigator(s), and the investigators are responsible for providing the up-to-date IB to the responsible IRBs/IECs. In the case of an investigator sponsored trial, the sponsor-investigator should determine whether a brochure is available from the commercial manufacturer. If the investigational product is provided by the sponsor-investigator, then he or she should provide the necessary information to the trial personnel. In cases where preparation of a formal IB is impractical, the sponsor-investigator should provide, as a substitute, an expanded background information section in the trial protocol that contains the minimum current information described in this guideline.

The IB should contain the following sections, each with literature references where appropriate.

4.1   General Considerations

The IB should include the following:

4.1.1   Title Page

This should provide the sponsor's name, the identity of each investigational product (i.e., research number, chemical or approved generic name, and trade name(s) where legally permissible and desired by the sponsor), and the release date. It is also suggested that an edition number, and a reference to the number and date of the edition it supersedes, be provided.

4.1.2   Confidentiality Statement

The sponsor may wish to include a statement instructing the investigator/recipients to treat the IB as a confidential document for the sole information and use of the investigator's team and the IRB/IEC.

4.2   Contents of the IB

The IB should contain the following sections, each with literature references where appropriate:

4.2.1   Table of Contents

An example of the Table of Contents is given in Table 1.

Table 1 Contents of an Investigator's Brochure (Example)

–	Confidentiality Statement (optional). The sponsor may wish to include a statement instructing the investigator/recipients to treat the IB as a confidential document for the sole information and use of the investigator's team and the IRB/IEC.
–	Signature Page (optional)
1	Table of Contents
2	A brief summary (preferably not exceeding two pages) should be given, highlighting the significant physical, chemical, pharmaceutical, pharmacological, toxicological, pharmacokinetic, metabolic, and clinical information available that is relevant to the stage of clinical development of the investigational product.
3	Introductory statement. Describing the product, active ingredients, proposed indications, and the relationship of the product to existing products for similar indications, and the general clinical approach to be taken in the study
4	Physical, chemical, and pharmaceutical properties, and formulation
5	Nonclinical studies. The results of all relevant nonclinical pharmacology, toxicology, pharmacokinetic, and investigational product metabolism studies should be provided in summary form.
6	Nonclinical pharmacology. A summary of the pharmacological aspects of the investigational product and, where appropriate, its significant metabolites studied in animals, should be included.
7	Pharmacokinetics and product metabolism in animals. A summary of the pharmacokinetics and biological transformation and disposition of the investigational product in all species studied should be given.
8	Toxicology. A summary of the toxicological effects found in relevant studies conducted in different animal species should be described.
9	Effects in humans. A thorough discussion of the known effects of the investigational product(s) in humans should be provided, including information on pharmacokinetics, metabolism, pharmacodynamics, dose response, safety, efficacy, and other pharmacological activities. Where possible, a summary of each completed clinical trial should be provided.
10	Pharmacokinetics and product metabolism in humans. A summary of information on the pharmacokinetics of the investigational product(s) should be presented.
11	Safety and efficacy. A summary of information should be provided about the investigational product's/products' (including metabolites, where appropriate) safety, pharmacodynamics, efficacy, and dose response that were obtained from preceding trials in humans (healthy volunteers and/or patients).
12	Marketing experience. The IB should identify countries where the investigational product has been marketed or approved. Any significant information arising from the marketed use should be summarized (e.g., formulations, dosages, routes of administration, and adverse product reactions).
13	Summary of data and guidance for the investigator. This section should provide an overall discussion of the nonclinical and clinical data, and it should summarize the information from various sources on different aspects of the investigational product(s), wherever possible. In this way, the investigator can be provided with the most informative interpretation of the available data and with an assessment of the implications of the information for future clinical trials.

4.2.2 Summary

A brief summary (preferably not exceeding two pages) should be given, highlighting the significant physical, chemical, pharmaceutical, pharmacological, toxicological, pharmacokinetic, metabolic, and clinical information available that is relevant to the stage of clinical development of the investigational product.

4.2.3 Introduction

A brief introductory statement should be provided that contains the chemical name (and generic and trade name(s) when approved) of the investigational product(s), all active ingredients, the pharmacological class of the investigational product(s) and its expected position within this class (e.g., advantages), the rationale for performing research with the investigational product(s), and the anticipated prophylactic, therapeutic, or diagnostic indication(s). Finally, the introductory statement should provide the general approach to be followed in evaluating the investigational product.

4.2.4 Physical, Chemical, and Pharmaceutical Properties and Formulation

A description should be provided of the investigational product substance(s) (including the chemical and/or structural formula(e)), and a brief summary should be given of the relevant physical, chemical, and pharmaceutical properties.

To permit appropriate safety measures to be taken in the course of the trial, a description of the formulation(s) to be used, including excipients, should be provided and justified if clinically relevant. Instructions for the storage and handling of the dosage form(s) should also be given.

Any structural similarities to other known compounds should be mentioned.

4.2.5 Nonclinical Studies

4.2.5.1 Introduction

The results of all relevant nonclinical pharmacology, toxicology, pharmacokinetic, and investigational product metabolism studies should be provided in summary form. This summary should address the methodology used, the results, and a discussion of the relevance of the findings to the investigated therapeutic and the possible unfavorable and unintended effects in humans.

The information provided may include the following, as appropriate, if known/ available:

- Species tested
- Number and sex of animals in each group
- Unit dose (e.g., milligram/kilogram (mg/kg))
- Dose interval
- Route of administration
- Duration of dosing
- Information on systemic distribution
- Duration of post-exposure follow-up

- Results, including the following aspects:
  Nature and frequency of pharmacological or toxic effects
  Severity or intensity of pharmacological or toxic effects
  Time to onset of effects
  Reversibility of effects
  Duration of effects
  Dose response

Tabular format/listings should be used whenever possible to enhance the clarity of the presentation.

In addition, the following sections should discuss the most important findings from the studies, including the dose response of observed effects, the relevance to humans, and any aspects to be studied in humans. If applicable, the effective and nontoxic dose findings in the same animal species should be compared (i.e., the therapeutic index should be discussed). The relevance of this information to the proposed human dosing should be addressed. Whenever possible, comparisons should be made in terms of blood/tissue levels rather than on a mg/kg basis.

Nonclinical Pharmacology A summary of the pharmacological aspects of the investigational product and, where appropriate, its significant metabolites studied in animals, should be included. Such a summary should incorporate studies that assess potential therapeutic activity (e.g., efficacy models, receptor binding, and specificity) as well as those that assess safety (e.g., special studies to assess pharmacological actions other than the intended therapeutic effect(s)).

Pharmacokinetics and Product Metabolism in Animals A summary of the pharmacokinetics and biological transformation and disposition of the investigational product in all species studied should be given. The discussion of the findings should address the absorption and the local and systemic bioavailability of the investigational product and its metabolites, and their relationship to the pharmacological and toxicological findings in animal species.

Toxicology A summary of the toxicological effects found in relevant studies conducted in different animal species should be described under the following headings where appropriate:

Single dose
Repeated dose
Carcinogenicity special studies (e.g., irritancy and sensitization)
Reproductive toxicity
Genotoxicity (mutagenicity)

4.2.6   Effects in Humans

4.2.6.1   Introduction
A thorough discussion of the known effects of the investigational product(s) in humans should be provided, including information on pharmacokinetics, metabolism, pharmacodynamics, dose response, safety, efficacy, and other pharmacological activities.

Where possible, a summary of each completed clinical trial should be provided. Information should also be provided regarding results of any use of the investigational product(s) other than from in clinical trials, such as from experience during marketing.

Pharmacokinetics and Product Metabolism in Humans A summary of information on the pharmacokinetics of the investigational product(s) should be presented, including the following, if available:

Pharmacokinetics (including metabolism, as appropriate, and absorption, plasma protein binding, distribution, and elimination)
Bioavailability of the investigational product (absolute, where possible, and/or relative) using a reference dosage form
Population subgroups (e.g., gender, age, and impaired organ function)
Interactions (e.g., product–product interactions and effects of food)
Other pharmacokinetic data (e.g., results of population studies performed within clinical trial(s)

Safety and Efficacy A summary of information should be provided about the investigational product's/products' (including metabolites, where appropriate) safety, pharmacodynamics, efficacy, and dose response that were obtained from preceding trials in humans (healthy volunteers and/or patients). The implications of this information should be discussed. In cases where a number of clinical trials have been completed, the use of summaries of safety and efficacy across multiple trials by indications in subgroups may provide a clear presentation of the data. Tabular summaries of adverse drug reactions for all the clinical trials (including those for all the studied indications) would be useful. Important differences in adverse drug reaction patterns/incidences across indications or subgroups should be discussed.

The IB should provide a description of the possible risks and adverse drug reactions to be anticipated on the basis of prior experiences with the product under investigation and with related products. A description should also be provided of the precautions or special monitoring to be done as part of the investigational use of the product(s).

Marketing Experience The IB should identify countries where the investigational product has been marketed or approved. Any significant information arising from the marketed use should be summarized (e.g., formulations, dosages, routes of administration, and adverse product reactions). The IB should also identify all the countries where the investigational product did not receive approval/registration for marketing or was withdrawn from marketing/registration.

4.2.7 Summary of Data and Guidance for the Investigator

This section should provide an overall discussion of the nonclinical and clinical data, and it should summarize the information from various sources on different aspects of the investigational product(s), wherever possible. In this way, the investigator can be provided with the most informative interpretation of the available data and with an assessment of the implications of the information for future clinical trials.

Where appropriate, the published reports on related products should be discussed. This could help the investigator to anticipate adverse drug reactions or other problems in clinical trials.

The overall aim of this section is to provide the investigator with a clear understanding of the possible risks and adverse reactions, and of the specific tests, observations, and precautions that may be needed for a clinical trial. This understanding should be based on the available physical, chemical, pharmaceutical, pharmacological, toxicological, and clinical information on the investigational product(s). Guidance should also be provided to the clinical investigator on the recognition and treatment of possible overdose and drug reactions that is based on previous human experience and on the pharmacology of the investigational product.

5. The Informed Consent

Informed consent is an ongoing process that must occur *before* any clinical trial-related procedures are conducted. Is a process by which a subject voluntarily confirms his or her willingness to participate in a particular trial, after having been informed of all aspects of the trial that are relevant to the subject's decision to participate. Informed consent is documented by means of a written, signed, and dated informed consent form.

In obtaining and documenting informed consent, the investigator should comply with the applicable regulatory requirement(s), and he/she should adhere to GCP [11] and to the ethical principles that have their origin in the Declaration of Helsinki. Prior to the beginning of the trial, the investigator should have the IRB/IEC's written approval/favorable opinion of the written informed consent form and any other written information to be provided to subjects.

The written informed consent form and any other written information to be provided to subjects should be revised whenever important new information becomes available that may be relevant to the subject's consent. Any revised written informed consent form and written information should receive the IRB/IEC's approval/favorable opinion in advance of use. The subject or the subject's legally acceptable representative should be informed in a timely manner if new information becomes available that may be relevant to the subject's willingness to continue participation in the trial. The communication of this information should be documented.

Neither the investigator nor the trial staff should coerce or unduly influence a subject to participate or to continue to participate in a trial.

The investigator, or a person designated by the investigator, should fully inform the subject or, if the subject is unable to provide informed consent, the subject's legally acceptable representative, of all pertinent aspects of the trial including the written information and the approval/favorable opinion by the IRB/IEC.

The language used in the oral and written information about the trial, including the written informed consent form, should be as nontechnical as practical and should be understandable to the subject or the subject's legally acceptable representative and the impartial witness, where applicable. Before informed consent may be obtained, the investigator, or a person designated by the investigator, should provide the subject or the subject's legally acceptable representative enough time and opportunity to inquire about details of the trial and to decide whether or not to participate in the trial. All questions about the trial should be answered to the satisfaction of the subject or the subject's legally acceptable representative.

Prior to a subject's participation in the trial, the written informed consent form should be signed and personally dated by the subject or by the subject's legally acceptable representative, and by the person who conducted the informed consent discussion.

If a subject is unable to read or if a legally acceptable representative is unable to read, an impartial witness should be present during the entire informed consent discussion. After the written informed consent form and any other written information to be provided to subjects is read and explained to the subject or the subject's legally acceptable representative, and after the subject or the subject's legally acceptable representative has orally consented to the subject's participation in the trial and, if capable of doing so, has signed and personally dated the informed consent form, the witness should sign and personally date the consent form. By signing the consent form, the witness attests that the information in the consent form and any other written information was accurately explained to, and apparently understood by, the subject or the subject's legally acceptable representative, and that informed consent was freely given by the subject or the subject's legally acceptable representative.

5.1 Content of the Informed Consent Form

The explanation of the informed consent, the summary of product characteristics and any other written information provided to subjects should include the following information:

1. That the trial involves research
2. The purpose of the trial
3. The trial treatment(s) and the probability for random assignment to each treatment
4. The trial procedures to be followed, including all invasive procedures
5. The subject's responsibilities
6. Those aspects of the trial those are experimental
7. The reasonably foreseeable risks or inconveniences to the subject and, when applicable, to an embryo, fetus, or nursing infant
8. The reasonably expected benefits. When there is no intended clinical benefit to the subject, the subject should be made aware of this.
9. The alternative procedure(s) or course(s) of treatment that may be available to the subject, and their important potential benefits and risks
10. The compensation and/or treatment available to the subject in the event of trial-related injury
11. The anticipated prorated payment, if any, to the subject for participating in the trial
12. The anticipated expenses, if any, to the subject for participating in the trial
13. That the subject's participation in the trial is voluntary and that the subject may refuse to participate or withdraw from the trial, at any time, without penalty or loss of benefits to which the subject is otherwise entitled
14. That the monitor(s), the auditor(s), the IRB/IEC, and the regulatory authority(s) will be granted direct access to the subject's original medical records for verification of clinical trial procedures and/or data, without violating the confidentiality of the subject, to the extent permitted by the applicable laws and regulations and that, by signing a written informed consent form, the subject or the subject's legally acceptable representative is authorizing such access

15. That records identifying the subject will be kept confidential and, to the extent permitted by the applicable laws and/or regulations, will not be made publicly available. If the results of the trial are published, the subject's identity will remain confidential

16. That the subject or the subject's legally acceptable representative will be informed in a timely manner if information becomes available that may be relevant to the subject's willingness to continue participation in the trial

17. The person(s) to contact for further information regarding the trial and the rights of trial subjects, and whom to contact in the event of trial-related injury

18. The foreseeable circumstances and/or reasons under which the subject's participation in the trial may be terminated

19. The expected duration of the subject's participation in the trial

20. The approximate number of subjects involved in the trial.

Prior to participation in the trial, the subject or the subject's legally acceptable representative should receive a copy of the signed and dated written informed consent form and any other written information provided to the subjects. During a subject's participation in the trial, the subject or the subject's legally acceptable representative should receive a copy of the signed and dated consent form updates and a copy of any amendments to the written information provided to subjects.

When a clinical trial (therapeutic or nontherapeutic) includes subjects who can only be enrolled in the trial with the consent of the subject's legally acceptable representative (e.g., minors, or patients with severe dementia), the subject should be informed about the trial to the extent compatible with the subject's understanding and, if capable, the subject should sign and personally date the written informed consent.

6. Essential Documents for Clinical Trial

The documents designed by ICH as "essentials documents" are those that individually and collectively permit evaluation of the conduct of a study and the quality of the data produced, and make it easier for monitors, auditors, and inspectors to evaluate the conduct of a trial and the quality of the data produced. They serve to demonstrate the compliance of the investigator, sponsor, and monitor with the standards of GCP and with the applicable regulatory requirements.

Essential documents also have other important purposes. Filing essential documents correctly at the investigator/institution and sponsor sites can greatly assist the investigator, sponsor, and monitor to conduct a clinical trial successfully. These are the documents that are usually audited independently by the sponsor and inspected by the regulatory authority(s) as part of the process to confirm the validity of the trial conduct and the integrity of the data collected.

The various documents are grouped in the ICH GCP guidelines [11] in three sections, as Table 2 shows, according to the stage of the trial during which they will normally be generated: (1) before the clinical phase of the trial commences, (2) during the clinical conduct of the trial, and (3) after completion or termination of the trial.

It describes the purpose of each document and specifies where they must be filed, in the investigator/institution files, the sponsor files, or both. Some documents may be combined, provided that individual elements are easily identifiable.

Table 2 Essential Documents of a Clinical Trial

Title of Document	Purpose	Located in File of	
		Investigator/Institution	Sponsor
Essential Documents Required *before* the Clinical Trial Starts			
Investigator's brochure (IB)	To document that relevant and current scientific information about the investigational product has been provided to the investigator	X	X
Signed protocol and amendments, if any, and sample case report form (CRF)	To document investigator and sponsor agreement to the protocol amendments(s) and CRF	X	X
Information given to trial subject:			
• Informed consent form (including all applicable translations)	• To document the informed consent	X	X
• Any other written information	• To document that subjects will be given appropriate written information to support their ability to give fully informed consent	X	X
• Advertisement for subject recruitment (if used)	• To document that recruitment measures are appropriate and not coercive	X	X
Financial aspects of the trial	To document the financial agreement between the investigator/institution and the sponsor of the trial	X	X
Insurance statement*	To document that all compensation to subject(s) for trial-related injury will be available	X	X
Signed agreements between involved parties:	To document agreements		
• Investigator/institution and sponsor		X	X
• Investigator/institution and CRO		X	X*
• Sponsor and CRO		X	X
• Investigator/institution and authority(s)*			X

Continued

Table 2 Essential Documents of a Clinical Trial—cont'd

Title of Document	Purpose	Located in File of	
		Investigator/ Institution	Sponsor
Dated, documented approval/favorable opinion of IRB/IEC of the following: • Protocol and any amendment • CRF* • Informed consent form(s) • Any other information provided to the subject(s) • Advertisement for subject recruitment* • Subject compensation* • Any other documents given approval/opinion	To document that the trial has been subject to IRB/IEC review and given approval/ favorable opinion. To identify the version number and date of the document(s)	X	X
IRB/IEC committee composition	To document that the IRB/IEC is constituted in agreement with GCP	X	X
Regulatory authority(s) authorization/approval/ notification of protocol*	To document appropriate authorization/ approval/notification by the regulatory authority(s) has been obtained prior to initiation of the trial in compliance with the applicable regulatory requirement(s)	X	X
Curriculum vitae for new investigator(s) and/ or sub-investigator(s)	To document qualifications and eligibility to conduct the trial and/or provide medical supervision of subjects	X	X
Normal value(s)/range(s) for medical/labo- ratory/technical procedure(s) and/or test(s) included in the protocol	To document normal values and/or ranges of the tests	X	X
Medical/laboratory/technical procedures/tests: certification or accreditation or established quality control and/or external quality	To document competence of facility to perform required test(s), and support reliability of results	X*	X

Document	Purpose		
Sample of label(s) attached to investigational product container(s)	To document compliance with applicable labeling regulations and appropriateness of instructions provided to subjects		X
Instructions for handling of investigational product(s) and trial-related materials, if not included in IB	To document instructions needed to ensure proper storage, packing, dispensing, and disposition of investigational products and trial-related materials	X	X
Shipping records for investigational product(s) and trial-related materials	To document shipment dates, batch numbers, and method of shipment of investigational products(s) and trial-related materials. Allows tracking of product batch, review of shipping conditions, and accountability	X	X
Certificate(s) of analysis of investigational product(s) shipped	To document identify, purity, and strength of investigational product(s) to be used in the trial		X
Decoding procedures for blinded trials	To document how, in case of an emergency, identify of blinded investigational product can be revealed without breaking the blind for the remaining subjects' treatment	X	X (third party if applicable)
Master randomization list*	To document method for randomization of trial population		X (third party if applicable)
Pre-trial monitoring report	To document that the site is suitable for the trial		X
Trial initiation monitoring report	To document that the trial procedures were reviewed with the investigator and the investigators trial staff	X	X

Continued

Table 2 Essential Documents of a Clinical Trial—cont'd

Title of Document	Purpose	Located in File of	
		Investigator/ Institution	Sponsor
Essential Documents Required during the Clinical Conduct of the Trial			
Investigator's Brochure updates	To document that the investigator is informed in a timely manner of relevant information as it becomes available	X	X
Any revision to the following: • Protocol/amendment(s) and CRF • Informed consent form • Any other written information provided to subjects • Advertisement for subject recruitment*	To document revisions of these trial-related documents that take effect during trial	X	X
Dated, documented approval/favorable opinion of IRB/IEC of the following: • Protocol amendment(s) • Revision(s) of informed consent form, any other written information to be provided to the subject, advertisement for subject recruitment*, any other document given approval/opinion, continuing review of the trial*	To document that the amendment(s) and/or revision(s) have been subject to IRB/IEC review and were given approval/favorable opinion. To identify the version number and date of the document(s)	X	X
Regulatory authority(s) authorizations/approvals/notifications where required for protocol amendment(s) and other documents	To document compliance with applicable regulatory requirements	X*	X
Curriculum vitae and/or other relevant documents evidencing qualifications of investigator(s) and sub-investigator(s)	To document qualifications and eligibility to conduct the trial and/or provide medical supervision of subjects	X	X
Updates to normal value(s)/range(s) for medi-	To document normal values and ranges that are revised during the trial	X	X

Document	Purpose		
...cedures/test: certification or accreditation or established quality control and/or external quality assessment or other validation*	throughout the trail period	X	X
Documentation of investigational product(s) and trial-related material shipment	To document shipment dates, batch numbers, and method of shipment of investigational products(s) and trial-related materials. Allows tracking of product batch, review of shipping conditions, and accountability	X	X
Certificate(s) of analysis for new batches of investigational products	To document identify, purity, and strength of investigational product(s) to be used in the trial		X
Monitoring visit reports	To document site visits by, and finding of, the monitor		X
Relevant communications other than site visits: letters, meeting notes	To document any agreements or significant discussions regarding trial administration, protocol violations, trial conduct, adverse event (AE) reporting	X	X
Signed informed consent forms	To document that consent is obtained in accordance with GCP and protocol and dated prior to participation of each subject in trial. Also to document direct access permission	X	
Source documents	To document the existence of the subject and substantiate integrity of trial data collected. To include original documents related to trial, to medical treatment and history of subject	X	
Signed, dated, and completed CRFs	To document that the investigator or authorized member of the investigators staff confirms the observations recorded	X (copy)	X (original)
Documentation of CRF corrections	To document all changes/additions or corrections made to CRF after initial data were recorded	X (copy)	X (original)

Continued

Table 2 Essential Documents of a Clinical Trial—cont'd

Title of Document	Purpose	Located in File of	
		Investigator/ Institution	Sponsor
Notifications by originating investigator to sponsor of SAE and related reports	Notification by originating investigator to sponsor of serious adverse events (SAE) and related reports	X	X
Notifications by sponsor and/or investigator, where applicable, to regulatory author-ity(s) and IRB(s)/IEC(s) of unexpected serious drug reactions and of other safety information	Notifications by sponsor and/or investigator, where applicable, to regulatory authority(s) and IRB(s)/IEC(s) of unexpected serious drug reactions	X*	X
Notification by sponsor to investigators of safety information	Notification by sponsor to investigators of safety information	X	X
Interim or annual reports to IRB7IEC and authority(s)	Interim or annual reports provided to IRB/IEC and to authority(s)	X	X*
Subject screening log	To document identification of subjects who entered pre-trial screening	X	X*
Subject identification code list	To document that investigator/institution keeps a confidential list of names of all subjects allocated to trial numbers on enrolling in the trial. Allows investigators/institution to reveal identify of any subject	X	
Subject enrollment log	To document chronological enrollment of subjects to trial numbers	X	
Investigational products accountability at the site	To document that investigational product(s) have been used according to the protocol	X	X
Signature sheet	To document signatures and initials of all persons authorized to make entries and/or corrections on CRFs	X	X

Document	Purpose		
Record of retained body fluids/tissue sample*	To document location and identification of retained samples if assays need to be repeated	X	X
Essential Documents Required *after the Closure of the Clinical Trial*			
Investigational product(s) accountability site	To document that the investigational product(s) have been used according to the protocol. To document the final accountability of investigational product(s) received at the site, dispensed to subjects, returned by subjects, and returned to sponsor	X	X
Completed subject identification code list	To permit identification of all subjects enrolled in the trial in case follow-up is required. List should be kept in a confidential manner and for agreed upon time	X	
Audit certificate (if applicable)	To document that audit was performed		X
Final trial close-out monitoring report	To document that all the activities required for trial close-out are completed, and copies of essential documents are held in the appropriated files		X
Final report by investigator to IRB/IEC where required, and where applicable, to the regulatory authority(s)	To document completion of the trial	X	
Clinical study report	To document results and interpretation of trial	X (if applicable)	X

*= where required.

7. Clinical Trial Files

7.1 Trial Master File

The documentation related to the clinical trial is archived in the Trial Master File (TMF). TMF consists of the essential documents that evaluate the execution of a clinical trial and the quality of the data collected [21]. These should be established at the beginning of the trial, both at the investigator/institution site and the sponsor site. These essential documents must prove that the investigator and the sponsor comply with the principles and guidelines on GCP and all the applicable requirements.

The final close-out of a trial can only be completed when the sponsor (or the individual responsible) has reviewed both the investigator/institution files and the sponsor files and confirmed that all the necessary documents are in the appropriate files.

All of the documents addressed in this chapter may be subject to, and should be available for, audit by the sponsor's auditor and inspection by the regulatory authority(s). The TMF shall provide the basis for audits to be carried out by the auditor (independent from the sponsor) and the inspections by the corresponding authorities.

TMF contains all the documentation of a clinical trial. Figure 2 shows the TMF's organization. It is constituted by the Sponsor's File and the Investigator's File (there will be as many files as PIs involved in the trial).

The content of the essential documents must be adjusted based on the specific nature of each stage of the clinical trial. The additional guidelines on the content of such documents published by the European Commission must also be taken into account [18].

The sponsor and the investigator will keep the essential documents of each clinical trial for at least 25 years after the trial is completed or for a longer period if required by the applicable requirements [12]. Essential documents must be filed so that they are easily available for the corresponding authorities.

Nevertheless, the medical records of the trial subject must be kept in accordance with the national law of every member state.

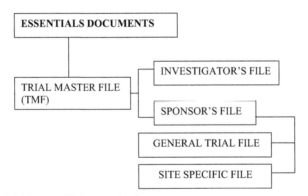

Figure 2 The Trial Master File's organization.

All changes to ownership data and documents must be documented. The new owner will assume the responsibilities of file-related tasks and keeping data.

The sponsor will designate the individual in charge of the files, which may only be accessed by the appointed individuals.

The formats used to keep the essential documents shall ensure that they remain complete and legible during the required retention period and that they are available to the corresponding authorities if required. Any modification of the records shall be traceable, thus the initial data, the corrected data, the date, and the signature of the author of the correction and/or update can be followed.

If a document is kept in more than one file, then the following apply:

1. There will be as many signed and dated original copies as needed.
2. The original document will be kept in the TMF, and copies of each of the investigator's files will be verified, signed, and dated.

To facilitate the management of the trial, audits, and inspections, the files will be organized according to an index and will be supervised following predetermined criteria.

An independent researcher who acts as sponsor must fulfill with the tasks of researcher and promoter. In this case, the file of the promoter and the researcher will be unified to avoid duplication.

The organization criteria and the indexing for the files shall be defined in the standard operational procedures of the sponsor. If the sponsor has not created the files, the organization criteria and the indexing for the files will be defined by the one performing the task (person or entity with the tasks transferred or outsourced).

A good manner is to classify the documents into sections identified by labeled tabs. The number of folders and their content will depend on the volume of documentation generated during the trial. Each folder should be identified with a side label.

The monitor (or CRA) is responsible for ensuring that all reviewed files contain all the essential documents and for document the issues that could have arisen. The closure of the trial can only be done once the sponsor (or the person responsible for it) has reviewed the investigator/institution's files and the sponsor's files and once they have confirmed that all the required documents are located in the appropriate files.

7.1.1 Sponsor's Files

A General Trial File and specific files for each trial site (Site Specific File) will be created (these folders are designed to keep a copy of the files of each investigator participating in the trial).

The Sponsor's File is created over the course of the trial. Throughout the trial, the documents will be properly filed in folders.

A good manner is to classify the documents into sections identified by labeled tabs. The number of folders and their content will depend on the volume of the documentation generated in the trial. Each folder will have an index describing the documents contained in each section.

7.1.1.1 General Trial File

The General Trial File contains all documents that are common to all participating trial sites (e.g., authorization by the corresponding Regulatory Agency, the approval of the corresponding IRB, amendments, signed annexes, insurance policy, protocols, IMP Dossier, IB).

When a document is to be kept in more than one file, the original will be kept in the General Trial File and copies of each investigator's file will be produced.

Table 3 shows an example of an index, and the documents classified in each section are attached below.

Table 3 **Index and the Documents Classified in Each Section of General Trial File (an Example)**

1. General documentation	1.1. Investigator's Brochure (all versions)
	1.2. Signed protocol
	1.3. Approved modifications
	1.4. Investigational medicinal product (IMP)
	1.5. Information given to subjects (all versions) and informed consent form (all versions)
2. IRBs	2.1. IRB resolution
	2.2. IRB composition
	2.3. Request for opinion on the IRB
	2.4. Clarification resolution correspondence to the IRB
	2.5. Request for relevant modifications (if applicable)
	2.6. Other communication with the IRB
3. Regulatory agency	3.1. Regulatory agency resolution
	3.2. Request for authorization to the regulatory agency
	3.3. EudraCT processing
	3.4. Requests for relevant modifications (if applicable)
	3.5. Clarification resolution correspondence to the regulatory agency
	3.6. Other communication with the regulatory agency
4. Agreements and contracts	4.1. Sponsor-CRO contract
	4.2. Sponsor confidentiality agreement to the CRO
	4.3. Representation authorization, sponsor to CRO
	4.4. Insurance policy/certificate
5. Investigational medicinal product	5.1. Instructions for handling of medicinal products
	5.2. Label models
	5.3. Unblinding procedures (if applicable)
6. Reports	6.1. Mid-term reports
	6.2. Annual reports
	6.3. Completed and simplified final reports
7. Others	7.1. General audit certificates (if any)
	7.2. Relevant correspondence
	7.3. Case report form (CRF)
	7.4. Others (please specify)

Each general folder may be identified with a side label containing the following:

The logo of the sponsor/individual or outsourced entity
Trial code
Folder number/total number of generated folders

7.1.1.2   Site Specific File
The Site Specific File contains the specific documentation of each of the trial sites (e.g., Principal Investigator's Commitment, financial agreement, CVs, normal laboratory values, correspondence with the PI and their team).

All original documents will be filed in the Site Specific File and, therefore, the documents from each site can be modified.

When a document must be kept in both the Site Specific File and the Investigator's File, the original document will be filed in the Site Specific File, and a copy will be filed in the Investigator's File, except for the Responsibilities Site Log and the Site Visit Log.

Table 4 shows an example of an index, and the documents classified in each section are attached below.

Each site specific folder may be identified with a side label containing the following:

The logo of the sponsor/individual or outsourced entity
Trial code
Site identification
Name of the PI of the site
Folder number/total number of generated folders

7.1.2   Investigator's File

The Investigator's File contains all the essential documents related to the clinical trial site. It is created over the course of the trial. Throughout the duration of the trial, the documents will be properly filed in folders.

The folder numbers and their content will depend on the volume of the documentation generated in the trial. Each folder will have an index with details of the documents in each section.

Each Investigator's File may be identified with a side label containing the following:

The logo of the sponsor/individual or outsourced entity
Trial code
Site identification
Name of the PI of the center
Folder number/total number of generated folders

A fundamental aspect about the investigator's file is that it presents two important sections for the development of the clinical investigation: the identification list of the subjects participating in the study and their original consent forms. The investigator is the only link between the sponsor and the regulatory authorities or between the Ethics Committee and the subjects participating in the trial. Due to this, and even though the participants have access to the original documents, the

Table 4 **Index and the Documents Classified in Each Section of Site Specific File (an Example)**

1. Agreements and contracts	1.1. Investigator's commitment
	1.2. Budget/financial agreement
	1.3. Contract with the site and/or the investigator(s)
	1.4. Approval by the managing department of the trial site
	1.5. Change of site report agreement
2. Investigators	2.1. CVs of the investigator and other key staff
	2.2. Responsibilities site log
	2.3. Suitability statement of the staff
	2.4. Suitability statement of the facilities
3. Laboratories	3.1. Range of normality in tests and upgrades
	3.2. Laboratory certificates/permits
4. Investigational product	4.1. Analysis certificates
	4.2. Reception of the medicinal product
	4.3. Medicinal product dispensing/administration
	4.4. Medicinal product destruction
	4.5. Site accountability log
5. Materials (if applicable)	5.1. Reception of materials
	5.2. Withdrawal of materials
6. Participating subjects	6.1. Screening patient log
	6.2. Randomized patient log
	6.3. Treatment assignation (if applicable)
	6.4. Unblinding envelopes (if applicable)
7. Monitoring	7.1. Minutes of investigators' meetings
	7.2. Monitoring planning
	7.3. Initial visit report
	7.4. Monitoring visit reports
	7.5. Closure visit report
	7.6. Site visit log
8. Adverse events and adverse reactions	8.1. Notifications of serious adverse events and reactions
	8.2. Follow-up and closure of serious adverse events and reactions
	8.3. Safety reports
9. Case report forms	9.1. Completed and signed CRFs (originals)
	9.2. Data related to screening failures
	9.3. Records of protocol deviations and waivers
	9.4. Correction and inconsistency (query) forms
10. Reports	10.1. Interim reports
	10.2. Annual reports
	10.3. Final report
11. Others	11.1. Audit certificate (if any)
	11.2. List of biological samples and transfers (if any)
	11.3. Documentation of changes in the file keeping
	11.4. Relevant correspondence
	11.5. Others (please specify)

investigator is responsible for safeguarding data confidentiality, and thus the information the investigator provides to the sponsor or the authorities should always be encrypted. Only the investigator should have the list of the participant codes of the study.

Similarly, the each patient's original medical records must remain with the rest of that patient's documents, and shall not be included in the sponsor's nor the investigator's file.

Table 5 show an example of an index, and the documents classified in each section are attached below.

Table 5 Index and the Documents Classified in Each Section of Investigator's File (an Example)

1. General documentation	1.1. Investigator's Brochure (all versions) 1.2. Signed protocol 1.3. Approved modifications 1.4. Investigational medicinal product (IMP) 1.5. Information given to subjects (all versions) and informed consent form (all versions)
2. IRB	2.1. IRB resolution 2.2. IRB composition 2.3. Request for opinion on the IRB 2.4. Clarification resolution correspondence to the IRB 2.5. Request for relevant modifications (if applicable) 2.6. Other communication with the IRB
3. Regulatory agency	3.1. Regulatory agency resolution 3.2. Request for authorization to the regulatory agency 3.3. EudraCT processing 3.4. Requests for relevant modifications (if applicable) 3.5. Clarification resolution correspondence to the regulatory agency 3.6. Other communication with the regulatory agency
4. Agreements and contracts	4.1. Sponsor-CRO contract 4.2. Sponsor confidentiality agreement to the CRO 4.3. Representation authorization, sponsor to CRO 4.4. Insurance policy/certificate 4.5. Approval by the managing department
5. Investigators	5.1. CVs of the investigator and other key staff 5.2. Responsibilities site log 5.3. Suitability statement of the staff 5.4. Suitability statement of the facilities 5.5. Relevant correspondence
6. Laboratories	6.1. Range of normality in tests and upgrades 6.2. Laboratory certificates/permits 6.3. Quality assurance programs (external/internal)

Continued

Table 5 **Index and the Documents Classified in Each Section of Investigator's File (an Example)—cont'd**

7. Investiga- tional medic- inal product	7.1. Instructions for handling of medicinal products 7.2. Analysis certificates 7.3. Reception of the medicinal product 7.4. Medicinal product dispensing/administration 7.5. Medicinal product destruction 7.6. Accountability log 7.7. Unblinding procedure (if applicable)
8. Materials	8.1. Reception of materials 8.2. Withdrawal of materials
9. Participating subjects	9.1. Screening patient log 9.2. Randomized patient log 9.3. Patient identification list 9.4. Signed informed consent forms
10. Monitoring	10.1. Initial visit report 10.2. Site visit log
11. Adverse events and adverse reactions	11.1. Notifications of serious adverse events and reactions 11.2. Follow-up and closure of serious adverse events and reactions 11.3. Safety reports
12. Case report forms	12.1. CRF model 12.2. Completed and signed CRFs (originals) 12.3. Data related to selection failures 12.4. Records of protocol deviations and waivers 12.5. Correction and inconsistency (query) forms
13. Reports	13.1. Interim reports 13.2. Annual reports 13.3. Final report (simplified and complete)
14. Others	14.1. Audit certificate (if any) 14.2. List of biological samples and transfers (if applicable) 14.3. Others (please specify)

7.2   *Period for Record Retention and Deletion of Files*

The essential documentation of noncommercial clinical trials must be kept for a minimum period of 25 years after the conclusion of the trial [13].

For the manufacturer's, investigator's, and sponsor's traceability records, the retention period of the essential documentation is 30 years after the last medicinal products released from a clinical trial.

The trial sites and PIs must receive information about these periods during the closing visit. In this visit, it will be noted that anyone who needs to check any essential document is entitled to a follow-up. The name, signature, start and end date of consultation will be recorded.

The storage conditions must permit the essential documents to be kept legibly and be accessible upon request of a regulatory authority. Any change in the location of the stored documentation must be recorded to allow tracking.

Adequate and suitable space shall be provided for the safe storage of all essential records of the studies conducted. The facilities must be safe, with appropriate controls and adequate protection from physical damage.

The storage of the sponsor's documentation may be transferred to a subcontractor (e.g., a commercial file), but the sponsor is ultimately responsible for the quality, integrity, confidentiality, and retrieval of the documents [11]. Directive 2005/28/EC [14] of Article 19 states, "The sponsor shall appoint individuals within its organization to be in charge of the files. Access to files shall be restricted to the named individuals in charge of them." The CROs should also follow this requirement. Withdrawal of essential documents from files should be under the control of the named individuals in charge of them (e.g., archive loans).

The investigator/institution is ultimately responsible for keeping the documents. If the investigator cannot maintain the responsibility for the essential documents (e.g., relocation, retirement), the sponsor should be notified in writing of this change and informed of to whom the responsibility has been transferred.

The sponsor shall obtain the investigator/institution agreement to keep the essential documents until the sponsor informs the investigator/institution that these documents are no longer required. The sponsor and the investigator/institution must sign the protocol, or an alternative document, to confirm this agreement.

The sponsor must notify the investigators in writing once the essential documents can be destroyed. The investigator/institution must take the necessary measures to prevent any accidents or the premature destruction of these documents.

Procedures that ensure the confidentiality of their content, such as prior shredding or specialized companies, will be used for the destruction of documents.

7.3 File Notes

Any incidents and discrepancies with respect to the filing method will be documented as "File Notes."

These documents will be filed in the section of the affected file according to the problem documented in it. These will be included in the investigator's files or in all of them if they are of general nature. They will be filed in the General Sponsor File if all sites are affected.

8. Sponsor' Study Audit and Inspections

One of the principles of ICH GCP is that all the information related to the clinical trial must be recorded, handled, and filed so that accurate reporting, interpretation, and verification can be carried out [11]. For that purpose, GCP establishes three methods for control and quality assurance: clinical trial monitoring, audits, and inspections.

The objectives of these monitoring processes are the following:

1. To protect the rights, safety, and welfare of subjects participating in clinical trials.
2. To verify the accuracy and reliability of data of clinical trials.
3. To assess compliance with regulations governing the conduct of clinical trials.

Adherence to the GCPs, including adequate human subject protection is universally recognized as a critical requirement to the conduct of research involving human subjects.

8.1 Definitions

A clinical trial inspection [11] is defined as an official review by the regulatory authority(s) of documents, facilities, records, and any other resources that are considered to be related to the clinical trial and that may be located at the trial site, at the sponsor's and/or CRO facilities, or at any other establishments deemed appropriate by the regulatory authority(s). It is important to note that the total or partial rejection of data provided by a research center may be the result of an inspection.

On the other hand, an audit [11] may be defined as an independent and systematic examination of the activities and documents related to a clinical trial to determine if the assessed activities related to the trial have been carried out and if the data is duly recorded, analyzed, and reported in accordance with the protocol, the SOPs of the sponsor, GCP, and the legislation in force.

Although it is true that these concepts are similar, the investigator staff usually tends to confuse them, which is why it is important to distinguish these processes in the following points:

Individual responsible for executing the audits/inspections: Inspections are performed by a corresponding regulatory authority, while audits are performed by personnel independent to the clinical trial, appointed by the sponsor.

Consequences: In the inspection report, the type of deviations and infringements are detailed. The infractions are classified as minor, severe, and very severe depending on the criteria of health risks. Each infraction will have a financial penalty, applying a grade of minimum, medium, and maximum to each level of infraction depending on the negligence and intentionality of the infringer or the fraud. However, in an audit, after detecting the findings, the sponsor has to implement a corrective measures plan, which can range from obtaining essential documents to the closure of a center participating in the trial or a complaint to the corresponding regulatory authorities.

8.2 Sponsor's Study Audit

The objective of an audit carried out by the sponsor, which is independent from regular monitoring or quality assurance functions, is to assess the performance of a trial and the fulfilment of the protocol, SOPs, GCP, and corresponding legal requirements.

8.2.1 Selection and Qualification of Auditors

The sponsor will appoint individuals who are independent of the system/clinical trials.

The sponsor will ensure that auditors are trained and have experience in conducting audits properly. The qualification of an auditor must be documented.

8.2.2   Audit Procedure

The sponsor shall ensure that clinical trial audits are carried out in accordance with the sponsor's written procedures (SOPs) about what and how to audit, the frequency of audits, as well as the format and content of the reports.

The sponsor's audit plan and the audit procedures of a trial shall be established according to the importance of the trial (based on the data to be submitted to the regulatory authorities), the number of trial subjects, the type and complexity of the trial, the level of risks for the trial subjects, and any other problems identified.

The observations and findings of the auditor must be documented. For this purpose, the auditor will create an audit report.

To preserve the independence and the value of audits, regulatory authorities should not request audit reports on a routine basis. The authorities can request access to an audit report for a particular case where there is evidence of a serious breach of GCP or during the course of a lawsuit.

When required by law or by the corresponding legislation, the sponsor must provide an audit certificate.

8.2.3   Recommendations

To carry out audits during the course of clinical trials [22] (when there are still few patients included), the following shall be accomplished:

To prepare the audit in advance, do not leave it until the last day
To collaborate with the auditor
The answers to all questions raised by the auditor must be documented
To answer only the questions raised
To be honest, to not hide anything
The research team must keep calm
Not to feel attacked or judged by the auditor
A perfect trial does not exist, and there is always room for improvement.

8.3   Inspections

As mentioned earlier, the main purpose of the inspections is to protect the rights and well-being of the subjects participating in a clinical trial and to ensure the reliability of the data collected from a clinical trial.

Inspections involve the evaluation of practices and procedures of both the sponsor and the investigator, or to determine compliance with the applicable regulations. There are several types of inspections:

- Random trial inspections: inspections carried out systematically in accordance with the "Plan for GCP Inspection" of the National Authority Competence. The objective is to verify that the data generated is reliable and that the clinical trial has been conducted according to the GCP principles and the sponsor's procedures.
- Inspection at the request of Marketing Authorization: inspections carried out at the request of a regulatory authority to clarify certain aspects of the performance of a clinical trial that is part of a marketing authorization request for a medicinal product.

When inspection occurs as a result of receiving a marketing application, including a comparison of the data submitted by the sponsor to the National Authority Competence with the source documents on the site of the clinical investigator (i.e., where data from the original source are recorded; also known supporting data), CRFs and files, clinical investigator in these cases, usually complete the study, possibly for a considerable time.

- Inspections due to complaints: Complaints may be presented by any entity participating in a clinical trial—the IRB, the Site or Institution, the subjects, or even the sponsor—if they consider that the corresponding legal requirements are not being fulfilled. If there is a "for cause" or inspection monitoring an ongoing study, the comparison of data usually involves only source documents and CRFs, as they cannot always be data supplied by the sponsor. Source documents may include medical notes, hospital records, laboratory reports, queries, etc.

Clinical trial inspections can take place as follows:

Before, during, or after the clinical trial
As a part of the verification process of the marketing authorization requests for a medicinal product
As follow-up to the suggested changes made in a previous inspection

Likewise, inspections may take place at the following:

The facilities of the sponsor and/or the CRO
The site
Any other establishment considered as appropriate for inspection by the corresponding authority, (IRB, hospital pharmacy, subjects, etc.)
The grading of each finding differs slightly depending on the National Authority Competence

The findings are classified by the GCP Inspectors of European Medicines Agency as "critical," "major," and "minor" according to the classification of GCP findings described in the "Procedure for reporting of GCP Inspections requested by the CHMP" [23,24]:

Critical:
- Conditions, practices, or processes that adversely affect the rights, safety, or well-being of the subjects and/or the quality and integrity of data.
- Critical observations are considered totally unacceptable.
- Possible consequences: Rejection of data and/or legal action is required.
- Remarks: Observations classified as critical may include a pattern of deviations classified as major, bad quality of the data, and/or absence of source documents. Manipulation and intentional misrepresentation of data belong to this group.
Major:
- Conditions, practices, or processes that might adversely affect the rights, safety, or well-being of the subjects and/or the quality and integrity of data.
- Major observations are serious findings and are direct violations of GCP principles.
- Possible consequences: Data may be rejected and/or legal action is required.
- Remarks: Observations classified as major may include a pattern of deviations and/or numerous minor observations.

Minor:
- Conditions, practices, or processes that would not be expected to adversely affect the rights, safety, or well-being of the subjects and/or the quality and integrity of data.
- Possible consequences: Observations classified as minor indicate the need for improvement of conditions, practices, and processes.
- Remarks: Many minor observations might indicate a bad quality, and the sum might be equal to a major finding with its consequences.

According to the FDA, the findings are classified as "No Action Indicated (NAI)," "Voluntary Action Indicated (VAI)," and "Official Action Indicated (OAI)":
- NAI: No objectionable conditions or practices were found during the inspection, or the objectionable conditions found do not justify any further regulatory action.
- VAI: Objectionable conditions or practices were found, but the District is not prepared to take or recommend any administrative or regulatory action.
- OAI: Regulatory and/or Administrative actions will be recommended.

An OAI classification may lead to a warning letter identifying serious deviations from regulations requiring prompt correction by the Investigator. A response is usually required within 15 days.

8.3.1 Inspection to the Sponsor

The sponsor is responsible for reaching an agreement with all the parties concerned that ensures direct access to all the trial sites, data/original documents, and the reports required to guarantee monitoring and audits by the sponsor and the inspections by the national and international regulatory authorities [25].

There are two types of inspections: those that generally inspect the sponsor and others specific to a clinical trial.

General inspections are employed to verify the following aspects:

Organization and staff: The sponsor must have enough trained personnel to perform trial-related activities.

Facilities and equipment: It will be verified that the sponsor's facilities are suitable for the filing and safekeeping of all the documentation, storing the samples, etc. Special attention must be paid to computer systems (hardware, software, reporting, etc.) to assess their validation and adaptation to the regulation requirements [21 CRF Part 11 (GAMP)].

Standard operating procedures (SOPs): During the inspection it will be verified that the sponsor has a control and quality assurance system with written SOPs to ensure that the trials are conducted and the data are generated, documented (recorded), and reported in accordance with the protocol, GCP, and the regulation in force.

Specific inspection of a clinical trial: The objective is to verify that the clinical trial has been conducted in accordance with GCP principles and that the data are generated, documented (recorded), and reported in accordance with the protocol, GCP, and the regulation in force.

8.3.2 Inspection in Site/Institution

As explained in GCP [26], both the investigator and the institution must allow the sponsor to carry out trial monitoring and audits and permit the corresponding authorities to carry out inspections. The refusal by the staff that has being inspected to facilitate the data/original documents and reports necessary for the correct performance of an inspection could be punishable.

In general, during an inspection in the site, the following aspects are usually verified:

Legal and administrative aspects: The approval of the corresponding IRB, the authorization by National Authority Competence, the approval of any relevant changes to the protocol, and the approval by the managing department of the trial site will be checked. The submission of the annual reports to the authorities and whether the trial has civil liability insurance shall be verified.

There should be a document related to duty transfer signed by the PI, the CVs of the research team, and a signed contract between the sponsor and the trial site.

All versions of the essential documents generated during the clinical trial protocol, the IB, the information for the patient, and the informed consent form, the CRF, etc., must be filed at the trial site.

It shall be verified that all subjects included in the clinical trial have given their consent in accordance with GCP.

To do this, it will be necessary to compare the data included in the CRF with the source document. Normally, the inspector selects a random sample of the trial variables for their verification.

It shall be verified that there is adequate traceability of the IMP from the point when it was received until it was dispensed or destructed, if applicable. If there is a double-blind trial where the emergency envelopes are retained by the pharmacy service, it shall be checked that these remain intact.

File of the trial documentation: It shall be checked that the storage conditions are appropriate and that the confidentiality of the subjects participating in the trial is upheld.

Monitoring: the number of the visits the monitor makes to the trial site, the date of the visits, and if these are in accordance with the monitoring plan will be examined. If an audit is carried out by the sponsor, the audit certificate must remain at the trial site.

8.4   Most Frequent Findings in Audits and Inspections

Absence of a list of collaborating investigators
Absence of documents showing the allocation of functions to the investigators
Lack of essential documents
Confidentiality problems
Disorganized files
Incomplete data in the CRF
Wrongly corrected data
Lack of evidence in the source document showing the inclusion of a subject in a clinical trial
Lack of evidence in the source document showing the test results
Lack of adherence to the protocol (subjects do not meet the selection criteria, visits were not carried out according to the work plan, etc.)
Incomplete data on the informed consent form or the signature was obtained after the inclusion in the trial
Deviations concerning the trial medicinal product (storage, labeling, etc.)

9. Conclusion

The path ahead to develop an ATMP, from laboratory to clinical practice, sometimes seems to be a hurdle race in which partial goals are being achieved to access the following stage of the development. The last of these stages is the clinical research with its own different phases. This path is full of regulatory requirements that will guarantee, on the one hand, the quality of any research carried out and, on the other hand, a risk–benefit balance positive for the patient when we come to the clinic.

Every effort made in the development of the medicinal product in research will be worthless unless the results coming from these trials are accepted by regulatory authorities of the countries involved. Consequently, any clinical trial, no matter its phase or nature, must be done according to the specific regulatory requirements. These requirements shall be not different when the development is carried out by a big pharmaceutical company or the clinical research is set up by small- or medium-sized biotechnological companies or nonprofit public institutions. And furthermore, regulations are more complex when they are related to ATMPs than to conventional medicines, even though, in most of these cell therapy products, its conception as medicinal product is still debatable.

The codes of GCP have been drawn up by most countries to ensure that the clinical trials and studies are performed in a scientific, humane, and ethical manner. The rights, safety, and well-being of the trial subjects are the most important considerations and should prevail over interests of science and society.

Take into account the limited number of ATMPs that have currently achieved to the clinic all over the world, the huge efforts that have been made by all the actors involved and the most of them have been assigned to the treatment of pathologies without efficient therapeutic alternatives, it seems necessary, without losing sight of the patient's protection, to make the regulatory requirements more flexible to facilitate the development of these specific medicinal products (or at least, keep them in line with the legal demands met by any research with other IMP different than ATMP), as they are the only future alternative for a part of the population.

References

[1] Maciulaitis R, D'Apote L, Buchanan A, Pioppo L, Schneider CK. Clinical development of advanced therapy medicinal products in Europe: evidence that regulators must be proactive. Mol Ther 2012;20(3):479–82.
[2] Welcome to the ICH Official Website [Internet]. International conference on harmonisation of technical requirements for registration of pharmaceuticals for human use expert working group (ICII). Geneva: ICII; 2014; URL: http://www.ich.org/home.html [accessed 27.02.15].
[3] Cuende N, Herrera C, Keating A. When the best is the enemy of the good: the case of bone-marrow mononuclear cells to treat ischemic syndromes. Haematologica March 2013;98(3):323–4.

[4] Cuende N, Zugaza JL. The andalusian initiative for advanced therapies. Eur Biopharm Rev 2010. Samedan Ltd, April issue.

[5] Cuende N, Izeta. Clinical translation of stem cell therapies: a bridgeable gap. Cell Stem Cell June 2010;6:508–12.

[6] Cuende N. Andalusian initiative for advanced therapies: fostering synergies. Stem Cells Transl Med April 2013;2(4):243–5.

[7] Als-Nielsen B, Kjaergard LL. Association of funding and conclusions in randomized drug trials. JAMA 2003;290:921–8.

[8] Trials of war criminals before the Nuremberg Military Tribunals under Control Council law no. 10, vol. 2. Washington, D.C.: U.S. Government Printing Office; 1949. pp. 181–182.

[9] U.S. Food and Drug Administration (FDA). The Belmont report ethical principles and guidelines for the protection of human subjects of research. The National Commission for the Protection of Human Subjects of Biomedical and Behavioural Research [Internet] Washington (DC): FDA; 1979. URL: http://www.fda.gov/ohrms/dockets/ac/05/briefing/2005-4178b_09_02_Belmont%20Report.pdf [accessed 27.02.15].

[10] World Medical Association (WMA). Declaration of Helsinki—ethical principles for medical research involving human subjects [Internet]. Ferney-Voltaire (FR): WMA; 2013. URL: http://www.wma.net/en/contact/index.html [accessed 27.02.15].

[11] European Medicines Agency (EMA). ICH Topic E 6 (R1) Guideline for Good Clinical Practice. Note for guidance on good clinical practice [Internet]. London: EMA; 2002. CPMP/ICH/135/95. URL: http://www.ema.europa.eu/docs/en_GB/document_library/Scientific_guideline/2009/09/WC500002874.pdf [accessed 27.02.15].

[12] Directive 2001/20/EC of the European parliament and of the Council of 4 April 2001 on the approximation of the laws, regulations and administrative provisions of the Member States relating to the implementation of good clinical practice in the conduct of clinical trials on medicinal products for human use. Off J Eur Communities May 1, 2001; L 1(21):34–44.

[13] Regulation (EU) N° 536/2014 of the European Parliament and the Council of 16 April 2014 on clinical trials on medicinal products for human use, and repealing Directive 2001/20/EC. Off J Eur Communities May 27, 2014; L 1(58):1–76.

[14] Commission Directive 2005/28/EC of 8 April 2005 laying down principles and detailed guidelines for good clinical practice as regards investigational medicinal products for human use, as well as the requirements for authorization of the manufacturing or importation of such products. Off J Eur Union April 9, 2005; L 91:13–9.

[15] European Commission. Eudralex, the rules governing medicinal products in the European Union. Detailed guidelines on good clinical practice specific to advanced therapy medicinal products [Internet], vol. 10. Brussels (BE): European Commission; 2009. URL: http://ec.europa.eu/health/files/eudralex/vol-10/2009_11_03_guideline.pdf [accessed 27.02.15].

[16] Directive 2004/27/EC of the European Parliament and of the Council of 31 March 2004 amending Directive 2001/83/EC on the Community code relating to medicinal products for human use. Off J Eur Union April 30, 2004;L 1(36):34–57.

[17] International Conference on Harmonisation of Technical Requirements for Registration of Pharmaceuticals for Human Use Expert Working Group (ICH). ICH guidelines [Internet]. Geneva; 1995. URL: http://www.ich.org/products/guidelines.html [accessed 27.02.15].

[18] U.S. Food and Drug Administration (FDA). Promoting safe and effective drugs for 100 years. Kefauver Harris Amendment [Internet]. Washington (DC): FDA; 2006. URL: http://www.fda.gov/AboutFDA/WhatWeDo/History/CentennialofFDA/CentennialEditionofFDAConsumer/ucm093787.htm [accessed 27.02.15].

[19] Centers for Disease Control and Prevention (CDC). The Tuskegee syphilis experiment. U.S. Public Health Service syphilis study at Tuskegee [Internet]. Atlanta (USA): CDC; 2013. URL: http://www.cdc.gov/tuskegee/index.html [accessed 27.02.15].

[20] International Conference on Harmonisation of Technical Requirements for Registration of Pharmaceuticals for Human Use Expert Working Group (ICH). Structure and content of clinical study reports- E3. Current step 4 version [Internet]. Geneva; 1995. URL: http://www.ich. org/fileadmin/Public_Web_Site/ICH_Products/Guidelines/Efficacy/E3/E3_Guideline.pdf [accessed 27.02.15].

[21] European Commission. Eudralex, the rules governing medicinal products in the European Union. Recommendation on the content of the Trial Master File and Archiving [Internet], vol. 10. Brussels (BE): European Commission; 2006. URL: http://ec.europa.eu/health/ files/eudralex/vol-10/v10_chap5_en.pdf [accessed 27.02.15].

[22] European Commission. Eudralex, the rules governing medicinal products in the European Union. Guidance for the preparation of good clinical practice inspections. European Commission [Internet], vol. 10. Brussels (BE): European Commission; 2008. URL: http://ec.europa.eu/health/files/eudralex/vol-10/chap4/guidance_for_the_preparation_of_ good_clinical_practice_inspection_reports_en.pdf [accessed 27.02.15].

[23] European Medicines Agency. Classification and analysis of the GCP inspection findings of GCP inspections conducted at the request of the CHMP [Internet]. London: EMA; 2014. EMA/INS/GCP/46309/2012. URL: http://www.ema.europa.eu/docs/en_GB/document_ library/Other/2014/12/WC500178525.pdf [accessed 27.02.15].

[24] European Medicines Agency. Procedure for reporting of GCP inspections requested by the CHMP [Internet]. London: EMA; 2013. EMA/INS/GCP/588734/2012. URL: http://www. ema.europa.eu/docs/en_GB/document_library/Regulatory_and_procedural_guideline/ 2009/10/WC500004479.pdf [accessed 27.02.15].

[25] European Commission. Eudralex, the rules governing medicinal products in the European Union. Annex IV to guidance for the conduct of GCP inspections—sponsor and CRO [Internet], vol. 10. Brussels (BE): European Commission; 2008. URL: http://ec.europa.eu/health/ files/eudralex/vol-10/chap4/annex_iv_to_guidance_for_the_conduct_of_gcp_inspections_-_sponsor_and_cro_en.pdf [accessed 27.02.15].

[26] European Commission. Eudralex, the rules governing medicinal products in the European Union. Annex I to guidance for the conduct of GCP inspections—investigator site [Internet], vol. 10. Brussels (BE): European Commission; 2008. URL: http://ec.europa.eu/health/ files/eudralex/vol-10/chap4/annex_i_to_guidance_for_the_conduct_of_gcp_inspections_-_investigator_site_en.pdf [accessed 27.02.15].

Glossary

Blinding/masking A procedure in which one or more parties of the trial are kept unaware of the treatment assignment(s). Single blinding usually refers to the subject(s) being unaware, and double blinding usually refers to the subject(s), investigator(s), monitor, and in some cases, data analyst(s) being unaware of the treatment assignment(s).

Case Report Form A printed, optical, or electronic document designed to record all of the protocol-required information to be reported to the sponsor on each trial subject.

Clinical trial/study Any investigation in human subjects intended to discover or verify the clinical, pharmacological, and/or pharmacodynamic effects of an investigational product(s), and/or to identify any adverse reactions to an investigational product(s), and/or to study absorption, distribution, metabolism, and excretion of an investigational product(s) with the object of ascertaining its safety and/or efficacy. The terms clinical trial and clinical study are synonymous.

Clinical trial/study report A written description of a trial/study of any therapeutic, prophylactic, or diagnostic agent conducted in human subjects, in which the clinical and statistical description, presentations, and analyses are fully integrated into a single report.

Contract research organization A person or an organization (commercial, academic, or other) contracted by the sponsor to perform one or more of a sponsor's trial-related duties and functions.

Essentials documents Those documents that individually and collectively permit evaluation of the conduct of a study and the quality of the data produced, make it easier for monitors, auditors, and inspectors to evaluate the conduct of a trial and the quality of the data produced.

Independent Ethics Committee An independent body (a review board or a committee, institutional, regional, national, or supranational), constituted of medical/scientific professionals and nonmedical/nonscientific members, whose responsibility it is to ensure the protection of the rights, safety, and well-being of human subjects involved in a trial and to provide public assurance of that protection by, among other things, reviewing and approving/providing favorable opinion on the trial protocol, the suitability of the investigator(s), facilities, and the methods and material to be used in obtaining and documenting informed consent of the trial subjects.

Informed consent A process by which a subject voluntarily confirms his or her willingness to participate in a particular trial, after having been informed of all aspects of the trial that are relevant to the subject's decision to participate. Informed consent is documented by means of a written, signed, and dated informed consent form.

Institutional Review Board An independent body constituted of medical, scientific, and nonscientific members, whose responsibility it is to ensure the protection of the rights, safety, and well-being of human subjects involved in a trial by, among other things, reviewing, approving, and providing continuing review of trials, of protocols and amendments, and of the methods and material to be used in obtaining and documenting informed consent of the trial subjects.

Investigator A person responsible for the conduct of the clinical trial at a trial site. If a trial is conducted by a team of individuals at a trial site, the investigator is the responsible leader of the team and may be called the Principal Investigator.

The Investigator's Brochure A compilation of the clinical and nonclinical data on the investigational product(s) that are relevant to the study of the product(s) in human subjects.

Monitoring The act of overseeing the progress of a clinical trial, and of ensuring that it is conducted, recorded, and reported in accordance with the protocol, SOPs, GCP, and the applicable regulatory requirement(s).

Randomization The process of assigning trial subjects to treatment or control groups using an element of chance to determine the assignments to reduce bias.

Source data All information in original records and certified copies of original records of clinical findings, observations, or other activities in a clinical trial necessary for the reconstruction and evaluation of the trial. Source data are contained in source documents (original records or certified copies).

Source documents Original document, data, and records (e.g., hospital records, clinical and office charts, laboratory notes, memoranda, subjects' diaries or evolution checklists, pharmacy dispensing records, recorded data from automated instruments, copies or transcriptions certified after verification as being accurate copies, microfiches, photographic negatives, microfilms or magnetic media, X-ray, subject files).

Sub-investigator Any individual member of the clinical trail team designated and supervised by the investigator at a trial site to perform clinical trial-related procedures and/or to make important trial-related decisions (e.g., associates, residents, research fellows).

List of Acronyms and Abbreviations

AIAT The Andalusian Initiative for Advanced Therapies
ATMPs Advanced Therapies Medicinal Products
CHMP Committee of Medicines of Human Use
CRA Clinical Research Associate
CRF Case Report Form
CRO Contract Research Organization
CV Curriculum Vitae
EMA European Medicines Agency
EU European Union
FDA Food and Drug Administration
GCP Good Clinical Practice
GMP Good Manufacturing Practice
IB Investigator's Brochure
ICH International Conference on Harmonisation
IEC Independent Ethics Committee
IMP Investigational Medicinal Product
IRB Institutional Review Board
NAI No Action Indicated
OAI Official Action Indicated
PI Principal Investigator
SOP Standard Operating Procedures
TMF Trial Master File
VAI Voluntary Action Indicated
WHO World Health Organization
WMA World Medical Association

Compatibility of GxP with Existing Cell Therapy Quality Standards

Roger Palau¹, Amy Lynnette Van Deusen²
¹Banc de Sang i Teixits, Edifici Dr. Frederic Duran i Jordà, Passeig Taulat, Barcelona, Spain;
²University of Virginia, Department of Pharmacology, Charlottesville, VA, USA

Chapter Outline

Academic institutions and transfusion centers are positioned to lead early-stage clinical development of cell-based advanced therapy medicinal products (ATMPs) [1]. Existing infrastructure, including blood and tissue bank facilities, translational medical programs, and clinical personnel experienced in transplantation, equips these noncommercial entities to quickly translate research-grade cell materials into safe and commercially viable products. Further, their unique expertise potentiates immediate and consistent supplies of human cells for a range of clinical uses, including induced pluripotent stem cells (iPSCs) generated from blood or other adult sources. In the future, generation of de novo hematopoietic stem cells (HSCs) or mature blood cells [2] could even put an end to bone marrow and blood supply shortages.

Before such potential can be realized, however, developers must negotiate the complex process of translating research materials into qualified clinical products suitable for human use. In noncommercial environments, current quality management structures for cell therapies are highly specific for blood and bone marrow-related products, including programs administered by The Joint Accreditation Committee-ISCT Europe and EMBT (JACIE, www.jacie.org), (NetCord, www.netcord.org), Foundation for the Accreditation of Cellular Therapy (FACT, www.factweb.org), and American Association of Blood Banks (AABB, www.aabb.org), or focused upon more global quality objectives, such as International Organization for Standardization (ISO, www.iso.org) guidelines (particularly ISO9001).

Further complicating the issue, good scientific practice (GxP) regulations—universally required for the development of new therapeutics—were originally designed for product development in corporate environments. While blood banks and transfusion centers have many processes easily adapted for compliance with certain features of GxP, other quality assurance (QA) aspects will require considerable effort. To their detriment, the majority of companies fail to grow quality systems in a coordinated and gradual manner, resulting in organizational divisions with a multitude of individualized practices and documentation. In this chapter, compatibility of current quality management schemes for blood banks and transfusion centers with GxP will be considered, as will impact of GxP implementation for noncommercial developers.

1. Quality Standards in Cell Therapy

Current quality standards for the collection, analysis, banking, and release of cells for transplantation or transfusion intend to promote both patient care and excellence in laboratory practice. However, the scope of global initiatives, such as the Alliance for Harmonisation of Cellular Therapy Accreditation (AHCTA, www.ahcta.org) and International Society for Cellular Therapy (ISCT, www.celltherapysociety.org), are heavily focused on blood and bone marrow-derived HSC-based therapies. ATMPs are viewed, more or less, as biologics by regulatory bodies as opposed to a transfusion product or organ transplant. This has specific implications for achieving marketing authorization, most notably the requirement for implementation of GxP quality processes.

1.1 Jurisdiction of Quality Systems

Accreditation of blood banks and transfusion centers is a voluntary process initiated by centers to demonstrate commitment to patient safety. For ATMPs, requirements for GxP implementation are determined through risk-based assessments that take into account source, ex vivo manipulation, and intended application of final products [3]. Accordingly, the majority of adult, embryonic, and iPSC-based therapeutics will be regulated as medical products because of the necessary ex vivo manipulation of cells [4]; however, this was not always true.

Prior to 2005, the United States Food and Drug Administration (US FDA) did not enforce regulations for "minimally manipulated" autologous adult stem cell therapies [5].

However, in 2010, the arguably vague phrase resulted in a legal battle between the FDA and a cell therapy manufacturer in Texas for a product requiring more than 24 h of processing. A successful verdict for the regulators [6] motivated the relocation of that manufacturer to Mexico [7] and publicly solidified the FDA's jurisdiction over "Human Cell, Tissue and Cellular and Tissue-Based Products." Applicable regulations are located in the eponymous US Federal Code of Regulations Title 21, Part 1271 [8].

In Europe, Regulation (EC) No 1394/2007 of the European Parliament established ATMPs as a new class in 2007, to include engineered or genetically manipulated cell therapies. These guidelines also established a Committee for Advanced Therapies within the European Medicines Agency (EMA) to review applications for new products in collaboration with the Agency's Committee for Medicinal Products for Human Use [9]. More recent reflection papers provide relevant information for quality and nonclinical development of cell-based medicinal products (CMBPs) [10] and use of stem cells in medicinal products [11]. (For the sake of clarity in this chapter, ATMPs will continue to refer to all advanced cell therapies unless specific regulations apply.)

1.2 Voluntary Accreditation Schemes

The routine use of blood and bone marrow products in medical practices across the globe requires significant infrastructure capable of safely handling materials. While registration and/or government licenses are almost always required for facilities hosting human cells or tissue, implementation of standardized quality systems, and accreditation are currently elective. Benefits of accreditation include external review of organizational structure and facilities, assistance with strategies for implementing quality systems, and continuing education about evolving regulations and standards.

Data suggests that accredited transplant programs have better patient outcomes, as evidenced by rate of relapse-free survival of patients after transplantation of allogeneic HSC progenitors. When a transplantation center was at an advanced phase of JACIE accreditation, survival rates were significantly higher, independent of year of transplantation and other risk factors [12]. As use of ATMPs grows and diversifies, entities like the ISCT are helping to refine standards and accreditation of cellular therapies on an international scale.

1.3 Mandatory Regulation under GxP

Rather than voluntary, GxP compliance is mandatory for ATMPs and entails confirmation of both safety and efficacy through extensive nonclinical and clinical studies. Briefly, GxP standards describe methods to ensure proper design, monitoring, and control of processes and environments. Compliance involves establishing strong quality management systems to document and maintain reliable facilities where trained personnel perform validated processes using qualified equipment through to final delivery of products according to good clinical practice (GCP) and as shown in Figure 1 [13,14].

Adherence to good manufacturing practice (GMP) assures the identity, potency, and purity of drug products by adequately controlling manufacturing operations to

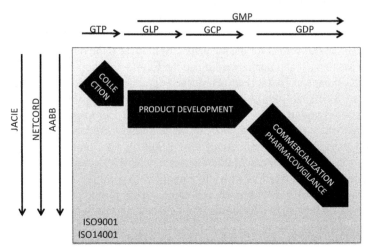

Figure 1 Quality standards in cellular therapies. This scheme shows the number of voluntary and mandatory quality assurance schemes converging in the development and commercialization of cell-based therapies.

prevent contamination, mix-ups, deviations, failures, and errors [15], while good tissue practice (GTP) is specifically intended to protect the integrity of collected human cells and tissue [16]. Additionally, all nonclinical studies used to justify investigational product characteristics or safety must be conducted under internationally agreed standards of good laboratory practice (GLP), which govern how studies are planned, performed, monitored, recorded, and archived.

During preclinical stages of development, emphasis is placed on proof-of-concept studies and prevention of communicable diseases within the laboratory through GLP [17,18] and GTP [19–21]. As clinical studies advance to Phase 1, all aspects of GxP must be implemented more rigorously. For this reason, every investigational new drug (IND, in the US)/Investigational Medicinal Product Dossier (IMPD, in Europe) application must include strategies for implementing GxP, along with an investigational plan, table of reagents, manufacturing and quality control (QC) data, nonclinical pharmacological/toxicological data, and clinical protocols [17,18]. Certifications such as JACIE, NetCord-FACT, and AABB facilitate development of investigational and clinical plans, although insufficiently for meeting GxP standards. For noncommercial developers, adapting systems inadequate for robustness rather than content is far easier than creating entirely new quality systems.

2. Adaptation of Existing Standards to GxP

Certification maps already present in blood, transfusion, and tissue banks include many processes that are easily adapted for compliance with certain features of GxP. Specific ISO9001, JACIE, FACT-NetCord, and AABB policies facilitate implementation of

GxP for staff management, metrological control of laboratory apparatus, and supervision of suppliers. However, other aspects of existing quality programs may require considerable restructuring or enhancement, including writing technical standard operating procedures (SOPs), training staff, examining reagent and drug product specifications, and performing GLP-compliant preclinical studies.

A comprehensive and systematic evaluation of all manufacturing facilities, processes, equipment, materials, and personnel to identify potential hazards is the first step in developing a robust GxP implementation strategy to mitigate or eliminate possible hazards [22]. For noncommercial ATMP developers, it is especially important to examine any existing certifications and quality systems to determine gaps and areas for improvement, since modification is generally more efficient that implementing entirely new quality plans. Specific aspects of individual certifications related to cell processing, which can be applied transversally across various certification schemes, are described throughout this section and diagrammed in Table 1.

- Quality management/policy
  Quality Review and Quality Plan can suit all the other standards because ISO 9001 covers all areas with specific standards.

  Using different policies as a starting point, each one relating to a specific quality standard, it is possible to generate a single quality policy for the whole company.

  A single annual quality plan is recommended. This is part of the quality review. In this document, each standard is discussed and takes into account their individual requirements. One quality management review program is capable of answering all the different standards. Integration of all aspects of all standards in one single document is recommended. If there are requirements that are very strict and specific for a given quality standard, then they could be included as appendixes.

- Risk management
  Risk management, mandatory for AABB and GxP, is being extended to the other standards. This is a good tool with applications to every critical aspect requiring modification or change.

- Quality audits/reviews/oversight
  Rather than audit plans for each standard, it is recommended to have a single plan approved by the general manager. This audit plan must be developed to address the specific points of every standard. Moreover, significant attention to personnel performing audits is recommended, as ongoing training and competence of each individual standard is required. A pool of auditors, comprised of individuals with different competences, is most effective in this regard.

- Equipment (including computer systems)
  It is recommended to create and maintain a single database regarding control of equipment. While this aspect varies widely according to each standard, GMP is one of the strictest, so more information for equipment present in GMP facilities must be included and maintenance performed (IQ/OQ/PQ in addition to calibration/verification).

- Documents and records
  To put in place a single document manager for the whole company, provided that some standards are very strict in having paper register of all documentation, the recommendation is to adopt the strictest regulation, in our case is governed by the GLP as opposed to ISO9001. Of course, this has an impact on the amount of documentation existing in the organization.

Table 1 Convergence of Quality Standards

	JACIE[a]	NETCORD[a]	AABB	GxP	ISO9001:2008
Quality management/policy	X	X	X	X	Quality review and quality plan can suit all the other standards
Risk management	N/A	N/A	Required and applicable from GxP	Systematic process for the assessment, control, communication, and review of risks to the quality of the medicinal product. It can be applied both proactively and retrospectively	N/A
Quality audits/reviews/oversight	X	X	X	X	Can cover all certifications with a single audit plan
Equipment (including computer systems)	X	X	X	Provided that DQ/IQ/OQ/PQ and validations are necessary, this fulfills the other standards from the technical point of view	Facilitates management of verifications/calibration plan for those noncritical equipment
Documents and records	X	X	X	Covers management, technical aspects, and particularly archive (particularly in GLP)	X

				Recommendations
Materials and reagents	X	X	X	Control guidelines can cover strictest standards JACIE, AABB, and GxP — X
Personnel	Very strict regarding qualification and re-qualification of personnel	X	X	X
CAPA	X	X	X	Can cover for all standards, including the documentation of root cause analysis and preventive and corrective actions — N/A
Control changes	N/A[b]	N/A[b]	N/A	Mandatory in GMP, this tool is applicable to all standards — N/A

Recommendations to simplify the quality management system when several certifications co-exist in an institution. X, requirement is applicable; N/A, Not applicable.

[a] A current EFI certificate for the laboratory is required.

[b] JACIE/Netcord: The terminology used is different than the one used in GxP. Instead of "Control of changes," JACIE/Netcord states that "changes to a process shall be verified or validated to ensure that they do not create an adverse impact anywhere in the operation." Therefore, these standards only apply to the validation or verification process that is part of "control of changes" under GMP.

- Materials and reagents
  Process for planning and controlling all steps in the acquisition and use of goods or supply items, a good elaboration of control guidelines is mandatory for a close control of materials and reagents. One must balance quality control of materials and reagents with the resources available for doing that job.
- Personnel
  To avoid each division qualifying their own personnel, it is proposed that a human resources department manages this quality aspect with a single system fulfilling all applicable standards.
- CAPA
  The main recommendation here is to generate a single database to manage "no conformities" and a QA unit that validates and follows up each event. All personnel must be trained in how to use relevant systems and databases for managing deviations and incidents.
- Control changes
  This is a very convenient tool for all quality management systems. This tool requires personnel involvement as changes must consider why, how, whom, etc. It has been accepted perfectly within the company.

2.1 Facilities and Equipment

Developers of ATMPs must demonstrate that manufacturing environments and equipment are adequately controlled for the generation of sterile products in accordance with GMP and relevant regulatory guidance. Facilities must include areas designated for sterile manufacturing and processing, as well as air-handling systems capable of preventing contamination and cross-contamination [23,24]. While all voluntary quality management schemes include similar policies for facilities and equipment contained within, both JACIE and FACT-NetCord fail to require specific measures for preventing adjustments to apparatuses that could impact product safety or quality [25].

This is especially important for GMP as, inevitably, a diverse range of equipment will be required to produce, analyze, and store cell products. Dedicated facilities and equipment should be employed whenever possible, and each instrument or metrological device must have appropriate installation qualification (IQ), operational qualification (OQ), and performance qualification (PQ) [15,26,27]. Further, it is strongly recommended that a single database be created to record environmental and apparatus controls, which should be adapted or integrated from any existing logs already required for ISO9001 compliance. Criteria for such records vary by quality standard, with GMP generally being strictest; thus, more information will be required in records, and equipment calibration must performed on an annual basis, or more frequently if recommended by the manufacturer.

Use of biosafety workstations for cell manipulations or other sterile processing (e.g., media preparation) must be documented to support control of GMP [28,29]. Further, sterile tasks should be performed under laminar airflow conditions meeting ISO Class 5 standards, which are related to the number and size of airborne particles allowed within a contained environment or cleanroom facility (ISO $5 = 100,000$ particles of $0.1 \mu m$ size per cubic meter) [30]. Biosafety cabinets, equipment providing temperature-appropriate storage, and other common laboratory instruments are already employed in blood banks and transfusion centers to store and test traditional cell therapy products. Assuming quality management protocols relevant to equipment certifications are in place, adaptation for ATMPs should only require strengthening of documentation systems for use and maintenance, as opposed to introduction of entirely new equipment and systems.

2.2 Personnel

Implementation of GxP standards has a substantial impact on personnel, mainly because all actions must be properly documented, reviewed, and archived, resulting in additional time for almost all tasks. Further, completion of appropriate training must be verified for all personnel involved in materials management, manufacturing, environmental monitoring, QC, QA, or any activity that has the potential to indirectly or directly affect product quality. While all voluntary accreditation schemes call for each employee to have defined qualifications and continuing education, GxP compliance involves additional training relevant to job duties and company quality policies.

Training records should be reviewed periodically to ensure training is still relevant to job responsibilities and up-to-date with current versions of SOPs. In addition to job-specific training, all personnel working in a facility manufacturing ATMPs should be familiar with the principles of relevant GxP quality systems [31,32]. To reduce the burden of these tasks, it is suggested that a single individual or department manages all personnel-related quality tasks. (How this relates to implementation of global quality systems will be discussed within the Section 2.5 of this chapter.)

2.3 Documentation

JACIE, FACT-NetCord, and AABB have thorough policies for documentation and control; however, requirements for GxP are even more stringent. First, more documentation is required, as even an SOP for writing SOPs is necessary to ensure sufficient content and format of all GxP-related documents. This indispensable document may already exist for facilities maintaining ISO9001 compliance, which are similar with regard to the extent of information that must be contained within documentation. Second, and equally important, systems for document review and change processes are more involved, requiring a more layered quality review approach. New procedures or changes to existing documents must be reviewed and approved by QA or another designated individual qualified to determine if changes are significant enough to require validation studies. All documents must be dated and a history of changes to standardized documents should be meticulously recorded. This "change control" process is designed to ensure consistent production of investigational products and validity of subsequently acquired clinical data [33,34].

Putting a single system and manager in place to oversee documentation activities is the best method to efficiently control the myriad of paperwork and review processes. A further step is to determine the most convenient aspects of each standard for the whole organization and adapt them to individual divisions accordingly. For example, control changes for GMP are a very valuable tool with applicability to a range of nonmanufacturing departments, including patient registration where appropriateness of informed consent documentation is always essential. Provided that some standards are very strict with regard to physical and digital versions of records, the general recommendation is to adopt the strictest regulation, frequently governed by GxP, rather than common blood-related or ISO9001 standards already implemented in blood banks and transfusion centers.

2.4 Materials

Sponsors of cell therapy products are responsible for ensuring the quality of every component used for production or packaging of materials. This means that all manufacturing constituents must be sterile, qualified for clinical use and/or removed from the final product in order to gain marketing authorization [35,36]. Management of materials begins with procurement, where appropriateness of suppliers and sources of material must be established [36,37]. The next step involves receiving materials, a process that includes inspection of package integrity and placement in storage at a proper temperature until QC can perform release testing [36,38].

JACIE and AABB have similar policies in place for inspection of materials that facilitate adoption of GMP and GLP. For example, prior to use in manufacturing, each lot must be evaluated for required documentation, to include a Certificate of Origin and Certificate of Analysis describing conformance to manufacturer's specifications [36,39]. For compliance with GMP, each sponsor must also generate specifications outlining acceptance criteria for qualifying each lot and aliquot safe for use in manufacturing. Additional SOPs defining handling, review, acceptance/rejection, and control of all materials used in manufacturing investigational product will likely also need modification. However, established relationships between blood banks/transfusion centers and medical supply vendors can expedite sourcing of GMP-compliant materials.

The principle component of any cell-based therapeutic is the cellular platform on which the therapy is based. Each ampoule of cells used to manufacture clinical products must have a recorded history, source, derivation, and characterization [40]. The origin and handling of cells is integral to regulatory risk assessments, as many potential dangers are linked to the health of the donor, quality of the derivation process, or subsequent storage procedures. Blood banks and transfusion centers are highly familiar with informed consent documentation, processes for donor testing and eligibility determination, which promotes adoption of similar GTP and GMP policies.

Potential cell sources for ATMPs include a patient's own (autologous) cells or cells obtained from another human donor (allogenic). While use of autologous cell sources reduces both the potential for negative immune responses and the burden of donor eligibility determination [41], allogeneic cell therapies use a consistent starting material, which may simplify manufacturing and QC. Prior to use in manufacturing, allogeneic cells must be tested for pathogens including cytomegalovirus (CMV), human immunodeficiency virus types 1 and 2 (HIV-1, -2), Epstein–Barr virus (EBV), Parvovirus B19, hepatitis B and C viruses (HBV and HCV), as well as adventitious agents such as bacteria, fungi, and mycoplasma [18,41–43]. Importantly, accredited blood banks and transfusion centers are equipped to handle such testing and processing of autologous cells and allogeneic products simultaneously, though modified labeling or storage procedures may be required for GxP.

In the allogeneic therapy setting, molecular analyses for determining donor–recipient compatibility are benefited by performance in EFI-certified laboratories, whose quality standards are compatible with JACIE and NetCord accreditation standards. It is the responsibility of product manufacturers to provide measures for identification and traceability of all materials used in the manufacture of investigational ATMPs, from initial receipt to use in individual batches [36,38]. Under JACIE and AABB accreditation schemes, requirements exist for a log containing the date of receipt, quantity, supplier's name, material lot number, storage conditions, and expiration date. Such logs must be rigorously kept for compliance with GMP and will likely need to be examined to ensure completeness and robustness. Finally, documentation must account for the location and environment of final cell products at all times, including during the shipment from manufacturing facilities to clinical sites [23,24,40], where compliance with GCP policies for receipt and administration of therapies begins to take effect.

2.5 Quality Management

Structures for quality management vary slightly between JACIE, FACT–FACT, and AABB accreditation schemes, but adherence to even the strictest of these guidelines may still be insufficient for GxP compliance. To improve the efficiency of adapting such processes, the QA unit should consider requirements of each certification to integrate a global plan for quality management. (If there are requirements that are very strict and specific for a given quality standard, they could be included as appendixes.) For GxP, the dynamic role of the QA unit involves examining all aspects of operations to support final product integrity, including the following:

1. Validating facilities and equipment were properly licensed, controlled, and functioning [44–46]
2. Verifying clinical product lots were produced and validated by qualified personnel
3. Confirming SOPs were adhered to by all personnel, at all times
4. Establishing aseptic technique and sterile processing were maintained
5. Authenticating QC data is sufficient for batch release
6. Endorsing completion of all relevant records, logs, and documentation
7. Authorizing release of final clinical product lots [47]
8. Investigating any nonconformance within SOPs, specifications, or facilities
9. Implementing corrective and preventative action (CAPA) for all incidents
10. Auditing all records, logs, and documentation for periodical revision and trending [48,49].

Ideally, a single quality system, with designated individuals who are each responsible for a certain aspect related to all certifications (i.e., documentation, metrological control of apparatus, CAPA, auditing, etc.), will result in a more organized, rational, and efficient methodology of work. For example, the individual responsible for enacting CAPA should have the capacity to handle an incident of nonconformity related to GMP and manage a separate nonconformity event related to ISO14001. Initially, extra training of personnel may be required, but this simplified arrangement accelerates all QA processes and reduces the amount of resources required.

To verify suitability for intended production of clinical-grade products, regulators will perform site inspections prior to initiation of Phase 1 clinical studies. Therefore, it is vital to have a quality unit with SOPs that designate individuals and schedules for performing internal audits to ensure facilities are ready for formal designation and certification. Similar to FACT-NetCord and AABB, but not JACIE standards [25], this quality unit must be independent of all product processing for GxP. These protections are put in place to ensure records are properly analyzed and reviewed by an unbiased individual, prior to batch release. The QA unit is also responsible for investigating deviations in SOPs or the failure of a batch or any of its components to meet specifications, regardless if the batch is distributed [48,49].

Rather than audit and CAPA procedures for each regulatory standard, it is strongly recommended to have a single plan in place, approved by the general manager. Further, since records of all investigations and CAPA must be coordinated between multiple departments, a single database of "nonconformities" should be generated to include a written record of investigations, conclusions, and summaries of follow-up actions. The QA unit must ensure follow-up, including any CAPA administered, is appropriate to the severity of the incident to safeguard similar events from impacting product quality in the future [50,51].

2.6 Logistics

In addition to product processing and quality, there are a multitude of logistical matters to consider when developing and commercializing ATMPs. For example, IND applications require descriptions of intermediate and final product packaging, as well as precautions to ensure the protection and integrity of products during use in clinical trials [52]. While full adherence to all facets of GxP may not be expected during early clinical stages, strategies for product packaging, labeling, storage, and transportation must be addressed within an IND application [22]. Fortunately, many existing products and regulatory guidelines for handling of human blood, cord blood, and bone marrow are readily adaptable for GxP compliance.

Conventions for identification and labeling of cell therapy products described within the International Standard for Blood and Transplant (ISBT) 128 Technical Specifications for Cellular Therapies [53] are supported by JACIE, AABB, and ISCT. ISBT standards are maintained by the International Council for Commonality in Blood Banking Automation (ICCBBA, www.iccbba.org), whose goal is to ensure medical products of human origin are assigned a unique identifier in a process standardized across international borders. This serves as a prime example of how the blood bank and transfusion industry is an archetype for the development of regulations for all advanced cell therapy products.

3. Impact of GxP Implementation

Every ATMP is unique in its origin, production, and intended use; therefore, GxP implementation and marketing authorization processes are highly individualized. For nearly all developers, there remain many technical and regulatory challenges to overcome before products are both suitable for the clinic and commercially viable. GxP implementation has a significant impact on a variety of aspects related to facility resources and frequently tests the business and financial acuity of developers. Early planning for such eventualities allows developers the time necessary to consider options for commercialization.

Qualifying and recording every element that comes into contact with product has a substantial impact on laboratory resources and finances. For example, each lot of product component, adjunct material, or final investigational product must be suitable for clinical use and tested for conformance with specifications regarding identity, sterility, purity, and potency [49,54]. Demonstrating this for any single item requires, at minimum, a combination of materials management, QC, and QA tasks. To ensure product integrity under GxP, increased documentation, trained personnel, as well as structured systems for cooperation between departments, will be required.

While accredited blood banks and transfusion centers have many policies and practices that are easily adapted to GxP, overlap between current standards commonly results in a multitude of certifications of all types (i.e., management, technical, etc.). Duplicated tasks have a negative impact on resources and are easily revealed if one considers the content of each standard: quality manual, SOP, management of apparatus, audits, and documentation. Clearly, the fact that different certifications have been achieved independently one from each other results in inefficiencies as shown in Table 1, especially for development programs with limited resources.

The financial aspects of implementing GxP quality systems are significant as multiple laboratory resources must be increased in size or be made more robust to support such extensive quality requirements. For example, performance of GLP studies and manufacture of GMP products to support preclinical safety and Phase 1 clinical trials is an expensive and time-consuming process. In addition to available grants and private investment, developers might consider partnership with a contract manufacturing (CMO) or research organization (CRO) to finance early cell therapy development and support investment at later stages.

4. Quality by Design

As with all clinical regulations, expectations for quality increase as a trial progresses and the number of potential patients increases. Taking future needs into consideration as early as possible will assist in development of scalable systems for Quality by Design (QbD)—a concept relating quality of product design, materials, and manufacturing processes to clinical performance of therapeutics as diagrammed in Figure 2 [55]. The cyclical nature of QbD processes reflects the continual optimization of products that occurs as clinical studies progress.

Strategies to implement QbD and GxP should be individualized for the facilities and personnel available during early clinical phases, then adapted to accommodate increased production during later stages. For example, a comprehensive strategy for examining, qualifying, and releasing materials for use in manufacturing should be developed early to provide consistent oversight of these functions as production increases, even if it requires more personnel. Once different policies from all required quality standards are considered in the context of available resources and future needs, it is possible to generate sustainable policies and scalable processes.

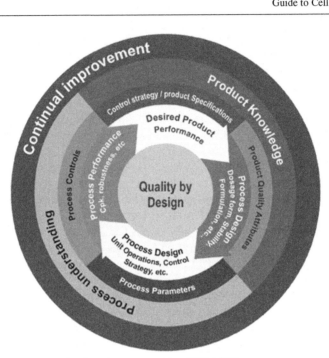

Figure 2 The cyclical nature of Quality by Design (QbD) [55].

5. Recommendations for Optimizing Integration of QA Systems

Developing an ATMP within the noncommercial environment of blood bank and transfusion centers can impart experience and knowledge applicable to all types of cell therapy product development. Further, quality practices currently implemented in these environments serve as templates for all future cell therapies, including those generated from adult and pluripotent stem cell sources. The complex nature of producing ATMPs requires a diverse set of expertise and an effective QA program to efficiently address regulatory, safety, and quality concerns. From our experience, creation of a single quality system incorporating the highest standard required is the most efficient use of resources, while early implementation of GxP-compliant methods advances clinical translation.

5.1 Early Adoption of GxP Systems Facilitates Product Commercialization

Documentation is required to demonstrate compliance for any quality management system and is key for establishing due diligence or intellectual property, therefore documentation strategies should focus on meeting the highest standards possible. Since quality systems already implemented by accredited blood bank and transfusion centers are thorough, taking into account nearly all aspects of GxP, the difference is often changes to key documentation impacting how practices are performed. For example, a document containing specifications for a particular product component may only require an extra field for one verification step required for GMP, but not AABB, JACIE, or ISO9001. Adding a single field to the SOP for preparing that specification represents a simple document change; however, training of personnel in how to properly complete this document is required, as is QA verification of all new documents, training, and processes. This one seemingly simple change requires a cascade of activities by multiple personnel to implement, but early adoption will allow impact on resources to be spread out over a longer period of time.

Considering strategies for enhancing existing quality systems at early time points also allows product developers to implement changes in a measured and progressive manner. Gradually making processes compliant with GxP will not only ensure facilities, equipment, and processes are suitable for future production of GMP products, it will allow personnel to provide feedback for optimization that can save valuable resources and time. While existing quality processes must be adapted for available resources, making small changes over time within an organization is almost always preferable to an overhaul of fundamental operating systems.

5.2 A Single Quality Management System is Most Efficient

While each department is required to successfully achieve quality standards and certifications related to their activities, working cooperatively reduces task duplicities and ensures consistent application of quality programs. Further, a single quality management system enables the QA unit to appropriately support all certifications, without having to perform the following:

- Administer different audit plans according to each quality standard
- Enforce different quality policies for each task or department
- Constantly revise multiple quality systems to satisfy individual quality standards
- Consolidate different systems of SOP
- Risk mixing up, for instance, "nonconformities" and "incidences"
- Investigate, perform, and report single CAPA according to multiple standard formats.

6. Conclusions

While blood banks, transfusion centers, and academic facilities are leading the way for research, most ATMPs remain in developmental phases. Translating the potential of cell therapies into the clinic requires strategies both for regulatory approval and commercialization of products that can feasibly be delivered to sites across the globe. To make that transition, noncommercial entities must consider long-term needs early in the development process, as well as consult with appropriate regulators and contractors for support.

An abundance of resources are available for ATMP developers, both from regulatory bodies, like the FDA and EMA, and nonprofit associations vested in the potential of cell therapies. In the United States, the FDA's Center for Biologics Evaluation and Research offers many different forms of assistance, including consultation with a variety of scientific and medical professionals. In Europe, the EMA's Committee for Medicinal Products for Human Use (CHMP) provides similar assistance under the recommendations of the Scientific Advice Working Party. Since every ATMP is unique in its origin, production, and intended use, pre-IND meetings frequently entail lengthy dialogues between researchers and regulatory officials to determine the most appropriate course of action. During early phases of development, support is focused on ensuring preclinical experimental design and analysis are sufficient. At later phases, assistance with design and execution of clinical experiments is provided according to regulatory frameworks outlined by The International Conference on Harmonisation of Technical Requirements for Registration of Pharmaceuticals for Human Use (ICH).

US and European regulatory frameworks call for coordination between regulatory committees and make reference to numerous published guidelines for ATMPs, however, they do not provide for a single codified quality system for the development of cell therapies. Rather than looking to each certification separately, product developers and manufacturers in all environments would benefit from a single system incorporating the best tools present in each quality standard for the whole company. Parallel to this, the improvement of quality standards in traditional blood and HSC-based products would encourage broader adoption of GxP. In noncommercial environments, quality management structures for cell therapies need adaptation to be less specific for blood and bone marrow-related products, and more open to the future of all advanced cell-based therapies.

References

[1] Maciulaitis R, D'Apote L, Buchanan A, Pioppo L, Schneider CK. Clinical development of advanced therapy medicinal products in Europe: evidence that regulators must be proactive. Mol Ther 2012;20:479–82.

[2] Yokoyama Y. Hematopoietic stem cells and mature blood cells from pluripotent stem cells. Nihon Rinsho Jpn 2011;69(12):2137–41.

[3] EMA Draft Guideline EMA/CAT/CPWP/686637/2011 – guideline on the risk-based approach according to annex I, part IV of Directive 2001/83/EC applied to advanced therapy medicinal products. February 11, 2013. Available from: http://www.ema.europa.eu/docs/en_GB/document_library/Scientific_guideline/2013/03/WC500139748.pdf [cited 01.03.15].

[4] US National Archives and Records Administration. Code of Federal Regulations. Minimal manipulation means. 2013: Title 21, Part 1271.3.

[5] US National Archives and Records Administration. Code of federal regulations. Scope 2013. Title 21, Part 1270.1(c).

[6] DeFrancesco L. FDA prevails in stem cell trial. Nat Biotechnol 2012;30(10):906.

[7] Cyranoski D. Controversial stem-cell company moves treatment out of the United States. Nat News January 20, 2013. Available from: http://www.nature.com/news/controversial-stem-cell-company-moves-treatment-out-of-the-united-states-1.12332 [cited 01.03.15].

[8] US National Archives and Records Administration. Code of federal regulations. Hum Cells, Tissues, Cell Tissue-Based Prod 2013. Title 21, Part 1271.

[9] EMA Regulation (EC) No 1394/2007, Advanced Therapy Medicinal Products (amending Directive 2001/83/EC, 6 Nov 2001, O.J. L 311 and Regulation (EC) 726/2004, 31 Mar 2004, O.J. L 136). Off J Eur Union November 13, 2007; L324. Available from: http://ec.europa.eu/health/files/eudralex/vol-1/reg_2007_1394/reg_2007_1394_en.pdf [cited 01.03.15].

[10] EMA draft guideline EMEA/CHMP/410869/2006 – guideline on human cell-based medicinal products. May 21, 2008. Available from: http://www.ema.europa.eu/docs/en_GB/document_library/Scientific_guideline/2009/09/WC500003898.pdf [cited 27.02.15].

[11] EMA guideline EMA/CAT/573420/2009 – reflection paper on clinical aspects related to tissue engineered products. September 19, 2009. Available from: http://www.ema.europa.eu/docs/en_GB/document_library/Scientific_guideline/2014/12/WC500178891.pdf [cited 01.03.15].

[12] Gratwohl A, Brad R, Niederwieser D, Baldomero H, et al. Introduction of a quality management system and outcome after hematopoietic stem-cell transplantation. J Clin Oncol May 20, 2011;29(15):1980–6.

[13] US National Archives and Records Administration. Code of federal regulations. Electron Rec Electron Signatures 2014. Title 21, Part 11.

[14] EMA Directive 2005/28/EC—on the wholesale distribution of medicinal products for human use. Off J Eur Union April 8, 2005; L91. Available from: http://ec.europa.eu/health/files/eudralex/vol-1/dir_2005_28/dir_2005_28_en.pdf.

[15] USFDA Center for Biologics Evaluation and Research. FDA guidance for industry: sterile drug products produced by aseptic processing – current good manufacturing practice. September 2004. Available from: http://www.fda.gov/downloads/drugs/.../Guidances/ucm070342.pdf [cited 01.03.15].

[16] EMA Directive 2003/94/EC - laying down the principles and guidelines of good manufacturing practice in respect of medicinal products for human use and investigational medicinal products for human use. Off J Eur Union October 2003; L262. Available from: http://ec.europa.eu/health/files/eudralex/vol-1/dir_2003_94/dir_2003_94_en.pdf [cited 01.03.15].

[17] US National Archives and Records Administration. Code of federal regulations. Curr Good Tissue Pract 2013. Title 21, Part 1271, Subpart D.

[18] EMA Directive 2004/23/EC – on setting standards of quality and safety for the donation, procurement, testing, processing, preservation, storage and distribution of human tissue and cells. Off J Eur Union November 13, 2007; L324. Available from: http://eur-lex.europa.eu/LexUriServ/LexUriServ.do?uri=OJ:L:2004:102:0048:0058:en:PDF [cited 01.03.15].

[19] US National Archives and Records Administration. Code of Federal Regulations. Good Laboratory Practice for Nonclinical Laboratory Studies. 2013: Title 21, Part 58.

[20] EMA Directive 2004/9/EC - on the inspection and verification of good laboratory practice (GLP). Off J Eur Union March 31, 2004; L102. Available from: http://eur-lex.europa.eu/LexUriServ/LexUriServ.do?uri=OJ:L:2004:050:0028:0043:EN:PDF [cited 01.03.15].

[21] EMA Directive 2004/10/EC - on the harmonisation of laws, regulations and administrative provisions relating to the application of the principles of good laboratory practice and the verification of their applications for tests on chemical substances. Off J Eur Union February 1, 2004; L50. Available from: http://eur-lex.europa.eu/LexUriServ/LexUriServ.do?uri=OJ:L:2004:050:0044:0059:EN:PDF [cited 01.03.15].

[22] US National Archives and Records Administration. Code of Federal Regulations. IND Content and Format. 2013: Title 21, Part 312.23.

[23] USFDA Center for Biologics Evaluation and Research. FDA guidance for industry: cGMP for Phase 1 investigational drugs. July 2008. Available from: http://www.fda.gov/downloads/Drugs/GuidanceComplianceRegulatoryInformation/Guidances/ucm070273.pdf [cited 01.03.15].

[24] EMA Guideline – Good Manufacturing Practice (GMP) guidelines. vol. 4. 25. Available from: http://ec.europa.eu/health/documents/eudralex/vol-4/index_en.htm [cited 01.03.15].

[25] Alliance for Harmonisation of Cellular Therapy Accreditation. Comparison of quality management. May 19, 2014. Available from: http://www.ahcta.org/docs/crosswalks/Quality%20Management%20Crosswalk%20version%20May%2019%202014.pdf [cited 01.03.15].

[26] US National Archives and Records Administration. Code of federal regulations. Process Validation 2013. Title 21, Part 820.75.

[27] ISO 13485. Medical devices – quality management systems – requirements for regulatory purposes. 2003. ISO 14644-4:2001. Available from: http://www.iso.org/iso/catalogue_detail?csnumber=36786.

[28] US National Archives and Records Administration. Code of federal regulations. Environ Control Monit 2013. Title 21, Part 1271.195.

[29] EMA guideline - manufacture of sterile medicinal products. Eudralex: vol. 4, Annex 1. November 25, 2008. Available from: http://ec.europa.eu/health/files/eudralex/vol-4/2008_11_25_gmp-an1_en.pdf [cited 01.03.15].

[30] Part 4: Design, construction and start-Up. Cleanrooms and associated controlled environments. ISO 14644–4:2001. Available from: http://www.iso.org/iso/catalogue_detail.htm?csnumber=25007.

[31] US National Archives and Records Administration. Code of federal regulations. Pers Qualif 2014. Title 21, Part 211.25.

[32] EMA guideline - personnel. Eudralex: vol. 4, Annex 2. February 16, 2014. Available from: http://ec.europa.eu/health/files/eudralex/vol-4/vol4-chap1_2013-01_en.pdf [cited 01.03.15].

[33] US National Archives and Records Administration. Code of federal regulations. Doc Controls 2013. Title 21, Part 820.40.

[34] EMA guideline - documentation. Eudralex: vol. 4, Part 1 [Chapter 4]. June 20, 2011. Available from: http://ec.europa.eu/health/files/eudralex/vol-4/chapter4_01-2011_en.pdf [cited 01.03.15].

[35] US National Archives and Records Administration. Code of federal regulations. Current good manufacturing practice in manufacturing, processing, packing, or holding of drugs. General 2013. Title 21, Part 210.

[36] EMA guideline - production. Eudralex: vol. 4, [Chapter 5]. December 6, 2005. Available from: http://ec.europa.eu/health/files/eudralex/vol-4/pdfs-en/2005_12_gmp_part1_chap8_en.pdf [cited 01.03.15].

[37] US National Archives and Records Administration. Code of federal regulations. Purch Controls 2013. Title 21, Part 820.50.

[38] US National Archives and Records Administration. Code of Federal Regulations. Receiving, in-process and finished device acceptance. 2015: Title 21, Part 820.80.

[39] US National Archives and Records Administration. Code of federal regulations. Supplies Reagents 2013. Title 21, Part 1271.210.

[40] USFDA Center for Biologics Evaluation and Research. Guidance for FDA reviewers and sponsors: content and review of chemistry, manufacturing and control (CMC) information for human somatic cell therapy investigational new drug applications. April 2008. Available from: http://www.fda.gov/BiologicsBloodVaccines/GuidanceComplianceRegulatory Information/Guidances/CellularandGeneTherapy/ucm072587.htm [cited 01.03.15].

[41] US National Archives and Records Administration. Code of federal regulations. Donor-eligibility determination not required. 2013. Title 21, Part 1271.90(a).

[42] US National Archives and Records Administration. Code of federal regulations. How do I screen a donor?. 2013. Title 21, Part 1271.75.

[43] US National Archives and Records Administration. Code of federal regulations. Gen Provis 2013. Title 21, Part 610, Subpart E.

[44] US National Archives and Records Administration. Code of federal regulations. Equipment 2013. Title 21, Part 1271.200.

[45] US National Archives and Records Administration. Code of federal regulations. Build Facil 2014. Title 21, Part 211.42.

[46] EMA guideline – premise and equipment. Eudralex: vol. 4, [Chapter 3]. August 13, 2014. Available from: http://ec.europa.eu/health/files/eudralex/vol-4/chapter_3.pdf [cited 01.03.15].

[47] US National Archives and Records Administration. Code of federal regulations. Build Facil 2014. Title 21, Subpart C.

[48] US National Archives and Records Administration. Code of federal regulations. Prod Rec Rev 2013. Title 21, Part 211.192.

[49] EMA guideline – pharmaceutical quality system. Eudralex: vol. 4, [Chapter 1]. January 31, 2013. Available from: http://ec.europa.eu/health/files/eudralex/vol-4/vol4-chap1_2013-01_en.pdf [cited 01.03.15].

[50] US National Archives and Records Administration. Code of Federal Regulations. Adverse reaction reports. 2013: Title 21, Part 1271.350(a).

[51] US National Archives and Records Administration. Code of Federal Regulations. Corrective and preventative action. 2013: Title 21, Part 820.100.

[52] USFDA Center for Biologics Evaluation and Research. Guidance for industry: container closure systems for packaging human drugs and biologics. May 1999. Available from: http://www.fda.gov/downloads/Drugs/Guidances/ucm070551.pdf.

[53] International standard for blood and transplant (ISBT) 128 technical specification for cellular therapies. Version 5.2.0. January 30, 2015. Available from: http://www.ICCBBA.org.

[54] US National Archives and Records Administration. Code of Federal Regulations. Testing and release for distribution. 2013: Title 21, Part 211.165.

[55] USFDA Department of Health and Human Services. Pharmaceutical quality for the 21st century: a risk-based approach progress report. May 2007. Appendix 19: Quality by Design Graphic. Available from: http://www.fda.gov/AboutFDA/CentersOffices/Officeof MericaProductsandTobacco/CDER/ucm128080.htm#APPENDIX19.

Glossary

American Association of Blood Banks (AABB) International nonprofit institution administering accreditation program for patient safety and quality management in facilities that handle or administer human blood, cord blood, bone marrow, or related cell therapy products.

Accreditation The process of meeting specific criteria (e.g., requirements for facilities, documentation, quality reviews, etc.) to obtain certification from a qualifying body that recognized standards have been met.

Alliance for Harmonisation of Cellular Therapy Accreditation (AHCTA) Organization formed by representatives of international cell therapy accreditation programs (see AABB, FACT, JACIE, NetCord, ISCT) whose purpose is to create a single set of quality, safety, and professional requirements for cell therapies.

Corrective and Preventative Action (CAPA) Processes or activities undertaken by an organization to eliminate causes of nonconformities, usually after a systematic investigation of the root cause for a specific deviation and identification of potential risks for recurrence. Specifically mentioned within GMP and ISO standards.

Cell-Based Medicinal Products (CBMP) Term used by European Medicines Agency to refer to (1) somatic cell therapies that are either substantially manipulated or intended for a use that is not related to the same essential function, or (2) engineered tissue or cell products.

Donor eligibility The process of determining if human donors of cell, tissue, or organs present any potential safety risks, specifically the transmission of communicable diseases or pathogens, prior to use of any donated material.

European Federation for Immunogenetics (EFI) International standards for the purpose of ensuring accurate and dependable histocompatibility testing consistent with the current state of technological procedures and the availability of reagents.

Good Scientific Practice (GxP) Generic term referring to quality standards or guidelines for GCP, GLP, GTP, and GCP. The general purpose of GxP is to ensure production of products that are safe for intended use through management of quality and traceability.

Good Clinical Practice (GCP) International quality standards for clinical trials using human subjects, provided by the International Conference on Harmonisation of Technical Requirements for Registration of Pharmaceuticals for Human Use (ICH).

Good Laboratory Practice (GLP) Set of principles for planning, executing, monitoring, recording, and archiving nonclinical studies assessing the safety or efficacy of potential medicinal products.

Good Manufacturing Practice (GMP) Set of established practices for producing food or drug products, for which conformance is mandatory. These extensive guidelines cover quality aspects of facilities, equipment, documentation, personnel, materials, suppliers, and review processes.

Good Tissue Practice (GTP) Guidelines for collection, handling, testing, documentation, and storage of human cells or tissue. Originally developed and enforced by the US FDA for HCT/Ps, these guidelines specifically exclude organ transplants and blood transfusions.

Foundation for the Accreditation of Cellular Therapy (FACT) International nonprofit institution establishing quality standards for medical and laboratory practice in cell therapies.

Food and Drug Administration (FDA) US Government federal agency empowered to regulate food and pharmaceutical products. It includes the Center for Biologics Evaluation and Research, which oversees HCT/Ps.

Human Cell, Tissue, and Cellular and Tissue-Based Product (HCT/P) Term used by US FDA to refer to (1) somatic cell therapies that are either substantially manipulated or intended for a use that is not related to the same essential function, or (2) engineered tissue or cell products.

International Council for Commonality in Blood Banking (ICCBBA) International organization administering standards related to blood banking, including global standards for identification and coding of medical products of human origin.

Installation Qualification (IQ) The process of establishing, though quantifiable and objective evidence, that all key aspects of equipment and ancillary systems for installation conform to manufacturer's specifications.

International Society for Blood Transfusion (ISBT) Scientific society promoting study and standardization of blood transfusion medicine, including issuance of ISBT 128 Standard Technical Specification for international labeling standards.

International Society for Cell Therapy (ISCT) Scientific society who mission is the advancement of cell therapies worldwide; provides education, consulting, and other resources for developers.

International Standardization Organization (ISO) Composed of representatives from various national standards organizations; publishes international standards, technical specifications, reports, and guides for a range of processes, including ISO9001 Quality Management Systems Requirements.

Joint Accreditation Committee-ISCT-EMBT (JACIE) International nonprofit institution administering accreditation program for patient safety and quality management in facilities that handle or administer cell therapy products; comprises representatives from ISCT and the European Society for Blood and Bone Marrow Transplantation (EMBT).

NetCord International foundation administering educational and accreditation programs for patient safety and quality management in facilities that handle or administer cord blood or related cell therapy products.

Operational Qualification (OQ) The process of establishing, by quantifiable and objective evidence, that process control limits meet all predetermined requirements during activities required for production.

Performance Qualification (PQ) The process of establishing, by quantifiable and objective evidence, that all processes performed (within reasonable and anticipated conditions) are capable of consistently producing product that meets predetermined specifications.

Quality Assurance (QA) Coordinated system of administrative and procedural activities designed to ensure quality requirements are consistently met.

Quality by Design (QbD) Principles developed by Joseph M. Juran used to advance product and process quality across a broad range of industries; adopted by US FDA for pharmaceutical and biological product development.

Quality Control (QC) The process by which all factors involved in a process (e.g., manufacture of a product) are reviewed for conformance to quantitative standards.

Quality management System of processes and activities implemented by an organization to ensure consistent product or service quality standards; generally includes planning, control, assurance, and improvement of all functions related to production.

Quality standard A document that provides a system of practices, specifications, guidelines, or requirements to ensure materials, products, processes, and services are fit for their specified function.

Standard Operating Procedure (SOP) Technical document providing detailed instructions for achieving uniformity within performance of a specific function.

Specification Technical document outlining a set of quality standards for materials, designs, products, or services.

Tissue Bank An establishment that collects, handles, and stores human cells or tissue; can be specialized (e.g., a Cord Blood Bank is a facility that stores human umbilical cord blood).

Transfusion Center An establishment that administers blood, bone marrow, or related component transfusions.

List of Abbreviations

AABB American Association of Blood Banks
AHCTA Alliance for Harmonisation of Cellular Therapy Accreditation
ATMPs Advanced Therapy Medicinal Products
CAPA Corrective/Preventive Action
CMBPs Cell-Based Medicinal Products
CMO Contract Manufacturing Organization
CRO Contract Research Organization
EFI European Federation for Immunogenetics
EMA European Medicines Agency
GxP Good Scientific Practice
GCP Good Clinical Practice
GLP Good Laboratory Practice
GMP Good Manufacturing Practice
GTP Good Tissue Practice
FACT Foundation for the Accreditation of Cellular Therapy

FDA Food and Drug Administration
HCT/P Human Cell, Tissue, and Cellular and Tissue-Based Product
HSCs Hematopoietic stem cells
ICH International Conference on Harmonisation
IQ/OQ/PQ Installation/Operational/Performance Qualification
IND Investigational New Drug
iPSCs Induced pluripotent stem cells
ISBT International Standard for Blood and Transplant
ISCT International Society for Cellular Therapy
ISO International Organization for Standardization
JACIE Joint Accreditation Committee-ISCT Europe
QA Quality Assurance
QbD Quality by Design
QC Quality Control
SOPs Standard Operating Procedures

Index

Printed in the United States
By Bookmasters